ELEMENTARY SCIENCE ACTIVITIES *for* ALL SEASONS

Julia Spencer Moutran, Ph.D.

illustrated by Shaela Cahill

THE CENTER FOR APPLIED
RESEARCH IN EDUCATION
West Nyack, New York 10995

10 9 8 7 6 5 4 3 2 1

Library of Congress Cataloging-in-Publication Data

Moutran, Julia Spencer
Elementary science activities for all seasons / Julia Spencer Moutran ; illustrated by Shaela Cahill.
p. cm.
ISBN 0-87628-302-4
1. Science—Study and teaching (Elementary) 2. Activity programs in education. I. Title.
LB1585.M68 1990
372.3′5044—dc20 89-17323
CIP

ISBN 0-87628-302-4

THE CENTER FOR APPLIED RESEARCH IN EDUCATION
BUSINESS & PROFESSIONAL DIVISION
A division of Simon & Schuster
West Nyack, New York 10995

Printed in the United States of America

DEDICATED

to

MY EDITOR,
Ann Leuthner, for
her belief in this book,

and to

MY FAMILY,
Alan, Meredith, and Melanie, for
their patience and love!

About the Author

Julia Spencer Moutran received her B.S. degree from the University of Virginia, and her M.A. and Ph.D. degrees from the University of Connecticut.

She has taught science in grades 3 through 8 in Tennessee and Connecticut, has served as an educational consultant, supervised elementary school teachers in science education, and coordinated Outdoor Education programs.

Dr. Moutran is also the author of two science storybooks, *The Story of Punxsutawney Phil, the Fearless Forecaster* and *Collecting Bugs and Things.*

About the Illustrator

A graduate of Williams College in Williamstown, Massachusetts, Shaela Cahill is an illustrator and graphic artist in New York City.

Ms. Cahill has taught art to children in Connecticut and at a summer camp in Maine.

About This Book

Dear Teachers,

Elementary Science Activities for All Seasons is divided into four sections, each representing the four climatic seasons: fall, winter, spring, and summer. Beginning with a fully illustrated picture of children outdoors during that season of study, each section has over twenty activities that include science experiments, crafts, songs, art, and teacher demonstrations.

A teacher's guide for each season offers suggestions for discussing the illustrated introductory picture, as well as ideas for using the unit activities in the classroom. Each section may be implemented as a unit of study for several months, correlated to the length of the season, or activities may be selected at random.

Since the activities are open ended, they may be used with students of different age and ability levels. All activity sheets are illustrated to appeal to children and stimulate their thinking.

A special feature of *Elementary Science Activities for All Seasons* is the numerous charts and graphs that help students record their conclusions and findings systematically. Included are opportunities for educational reward that motivate students to complete the activities. Distributing science certificates upon completion of each unit and using science flow charts will help students work enthusiastically to complete their activities.

Elementary Science Activities for All Seasons also lends itself to wonderful science displays, bulletin boards, and sharing with parents and other students in many enriching ways.

There are also many resources provided:

- ready-to-use activity sheets
- correlated science film and video lists

- information for ordering science lab materials
- ready-to-use science progress flow charts
- science certificates
- ready-to-use parent letters
- quick and easy-to-read teacher's guides
- exciting science crossword puzzles
- opportunities for homework and/or follow-up
- stimulating ideas and suggestions for teaching
- answer keys to activity sheets and crossword puzzles

All activities can be done in a variety of settings, such as small groups, large groups, demonstrations, home-study, or even nature clubs and summer school.

Written in clear, easy-to-read form, the activities are reproducible and "ready to use" with your students. All require a minimum amount of science equipment, so you do not have to spend hours preparing for activities and getting lab equipment ready. This will enhance the valuable time you spend with your students discussing the conclusions of the activities and stimulating their interests in future study.

I hope you will enjoy using *Elementary Science Activities for All Seasons*. It is designed to be used throughout the school year, taking advantage of both indoor and outdoor settings during the four seasons, regardless of where you live.

Julia Spencer Moutran

Science Season Certificate

This is to certify that ______________

has completed the required lessons

in the ____________ unit of

Science Activities for All Seasons.

______________ Parent's Signature ________ date

______________ Teacher's Signature

Name ____________________ Teacher ______________________

SEASON & LESSON NUMBER	DATE STARTED	DATE COMPLETED	SCORE

Section One: THE FALL SEASON • 1

Activities

The Fall Season

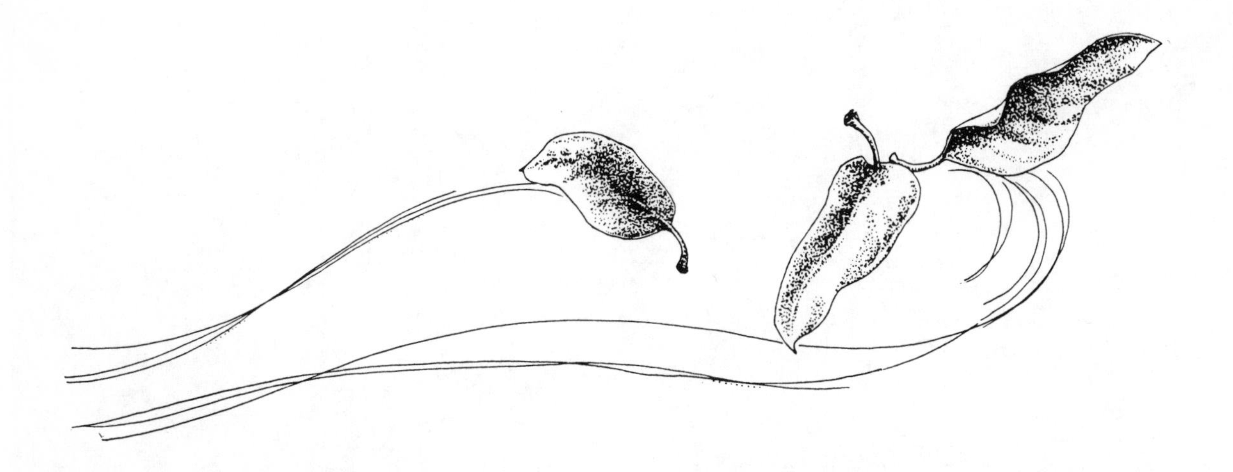

TEACHER'S GUIDE FOR FALL

Congratulations and good luck! You are about to begin a very exciting and interesting year in science education. This activity book will help you implement your science goals and curriculum.

To help your students become more organized and responsible for their own progress and activity completion, begin each of the four sections with the Science Progress Flow Charts. Reproduce the Science Progress Flow Charts and use them for recording student progress; give students a copy of their charts to help them see their progress and completion of the activities.

There are twelve activities in the Fall Season Section. All activities are numbered for the season; e.g., "F1" means "Fall Season, Activity 1."

After the twelve major activities (F1–F12), another section follows, FALL SEASON ENRICHMENT ACTIVITIES. These are activities that are more interdisciplinary in scope, bringing concepts of science into other subject areas, including language arts, mathematics, cooking, music, and arts and crafts. The concentration is on the application of science concepts to other areas of the curriculum.

After selecting the activities for your students, reproduce the activity sheets and bind them in a science notebook, using the full-page illustration on page 2 as the cover. Be sure to discuss the illustration in detail, covering these topics or observations:

- Geese are flying south.
- The sheep are growing heavy coats of wool after summer's shearing.
- The groundhog is preparing a burrow for hibernation.
- The squirrel is storing nuts in the tree.
- The garden and apples have been harvested.
- Leaves have fallen.
- The bulbs are dormant.
- Shadows are lengthening.
- Altocumulus clouds bring fall rains.
- The children's clothing indicates temperature changes.

For your audio-visual needs, check the Coronet/MTI list of films and science videos recommended for the unit. The list is found in Appendix 3 of this book.

Send a copy of the *Parent Letter* to your student's parents, explaining the activity sheets they will use. The letter is ready to be reproduced on your school stationery for an official look!

Use a variety of learning settings with these easy-to-use activity sheets. Set up an *activity* or *learning lab*. This mini-lab can be a quiet spot or corner of your classroom where the students can work independently and complete activity sheets on their own.

Plan a fun day in science, like a Fall Brunch. Invite a few guests and display your Science Notebooks. Serve some of the foods from "Science Cookery" in the Enrichment Section.

Whichever you select, enjoy the school year and have fun teaching science! You are providing your students with activities that will help them learn skills, concepts, and important information. You are also developing them as independent learners and inspiring them to seek more knowledge about the wonderful world of science.

Dear Parents,

This year, your child will be doing lots of exciting science experiments and activities. Many hands-on activities will come from a science book entitled *Elementary Science Activities for All Seasons*.

All the lessons in the book relate to the four climatic seasons and changes in the environment corresponding to the different seasons. Students will learn about growth, survival, and adaption of animals, insects, plants, and trees. They will study about weather, climate, gardening, constellations, and what causes the seasons to change.

The science lessons are designed to develop students' process skills in seven areas: Identification, Measurement, Observation, Description, Prediction, Analysis, and Application. Student Progress Charts are used to monitor student progress.

Some of the activities will be brought home for homework or follow-up. This will give you and other members of the family the chance to share in your child's learning experience.

Thank you for your continued support and interest in your child's progress in science.

Sincerely,

____

Teacher

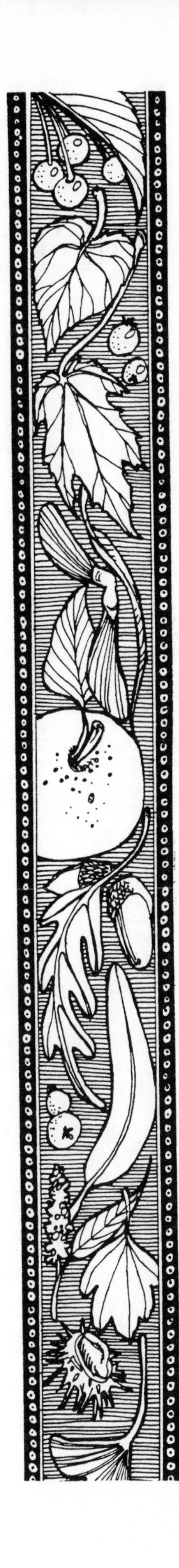

Name ______________________ Date ______________________

INTRODUCTION TO THE CLIMATIC SEASONS

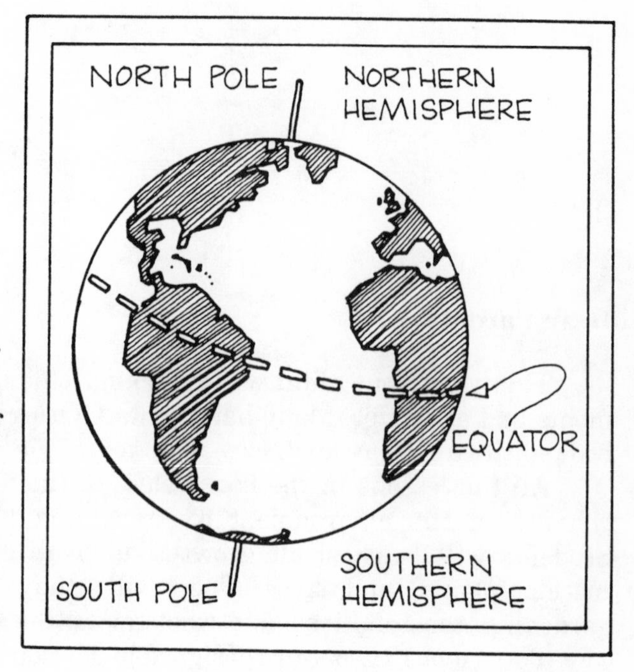

There are two to four seasons in the calendar year, depending on where you live. Seasons are determined by the changes in the amount of sunlight and precipitation the area receives. Each season is unique. Seasons vary in the length of time they last.

Here is a questionnaire for you to fill out. Think about where you live in relation to the equator on our globe.

1. How many seasons do you have where you live? ______________________
2. What is your favorite season? ______________________
3. Guess your teacher's favorite season. ______________________
4. Most seasons are how many months long? ______________________
5. Is this statement TRUE or FALSE? "The seasons of people living near the equator are not as distinctive as the seasons of people living away from the equator." ______________________
6. Is this statement TRUE or FALSE? "Near the North and South Poles, there appear to be two seasons, marked by periods of darkness and lightness." ______________________
7. Describe the trees and leaves and how they change during the fall season in your area. (Use the back of this paper.)

F1: What and When Is Fall?

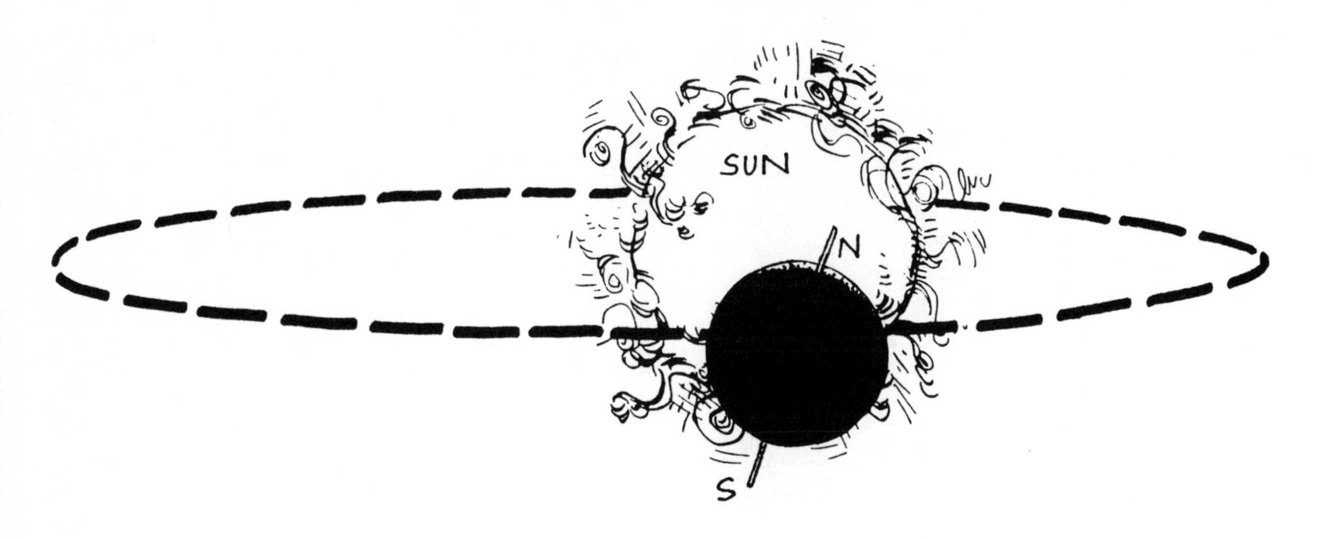

Use this activity at the beginning of the unit on fall. It will help students understand that the earth spins on its axis in its orbit around the sun. The earth completes one-quarter of its revolution between September 23 and the first day of winter, around December 21. The North Pole tilts away from the sun in the fall.

In this investigation the students will use a ball to represent the earth. Be sure they have it angled so the North Pole tilts away from the sun for one-quarter of its orbit.

The illustration on page 8 shows what the students will be acting out in the investigation, "What and When Is Fall?"

EQUIPMENT:

Globe
Flashlight, light, or clip light
Ball to represent the Earth, with North and South Poles marked
Pencil
Tape
Piece of string or rope about 12 feet long

DIRECTIONS:

1. Look at the globe and locate the North and South Poles and equator.
2. Mark the ball with the Poles and equator using the pencil.
3. The person holding the clip light or flashlight stands in the middle of an ellipse. The ellipse is made by taking the string or rope and placing it on the floor in the shape of an ellipse. Tape the ends of the string together.
4. The person holding the ball representing the earth walks along the rope in an ellipse pattern. Students observe the path of the earth. One quarter of the path represents a season. During fall, the earth must be turned so the North Pole tilts away from the sun.
5. The room is darkened and the sun shines as the clip light or flashlight is turned on. The earth moves in its orbit around the sun.
6. Discuss the different positions of the earth during the four seasons, as shown in the illustration on page 8.

HERE'S WHAT YOU'RE ACTING OUT:

THE EARTH IS TILTED ON ITS AXIS. IN OTHER WORDS, IF YOU WERE ABLE TO STICK A LONG POLE THROUGH THE EARTH, FROM THE NORTH POLE TO THE SOUTH POLE, IT WOULD LOOK LIKE THIS , AND NOT LIKE THIS . THIS TILTING OF THE EARTH CAUSES THE CHANGES IN TEMPERATURE AND WEATHER DURING THE DIFFERENT SEASONS OF THE YEAR. IT ALSO EXPLAINS WHY OUR DAYS ARE LONGER IN THE SUMMER AND SHORTER IN THE WINTER.

HERE'S HOW:

F2: Examining the Trees and Leaves

Take your class for an outdoor walk in the fall to observe the trees and leaves. They can complete the "Tree Observation Chart" while on the walk. They can also do this activity at home for follow-up or homework.

Remind the students that all seasons are good times to observe trees. In the United States, there are more than 800 different kinds of trees. Trees adapt to seasonal changes. Some change colors or drop their leaves in the fall.

Trees are categorized into three main areas: palm, conifer, and broadleaf. The largest family is the broadleaf, including many deciduous trees like the maple, birch, and oak.

On the walk remind children to think about the animals that make their homes in trees or depend upon the trees for survival during the winter.

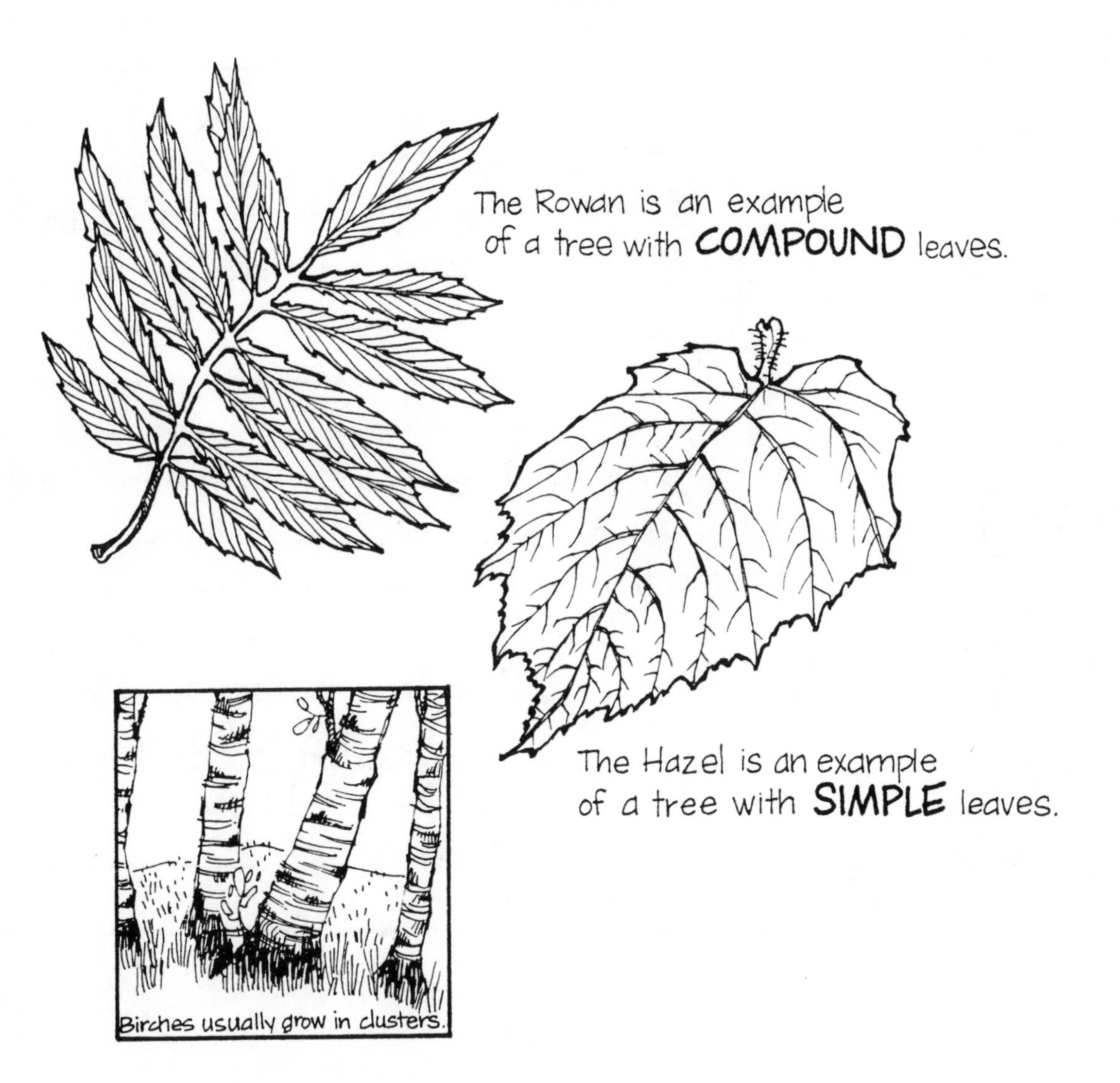

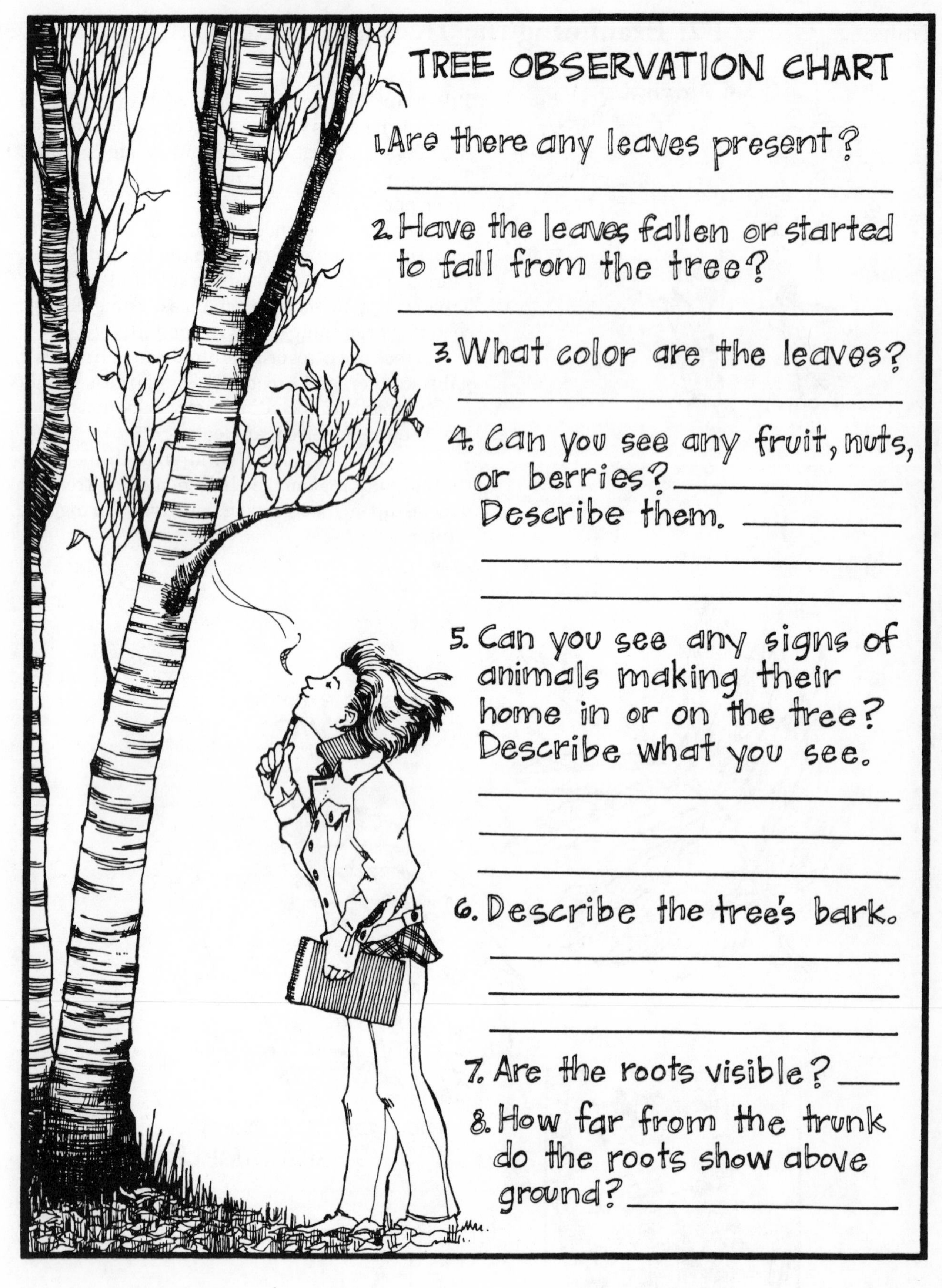

TREE OBSERVATION CHART

1. Are there any leaves present?
____

2. Have the leaves fallen or started to fall from the tree?
____

3. What color are the leaves?
____

4. Can you see any fruit, nuts, or berries? ______________
Describe them. ______________
____
____

5. Can you see any signs of animals making their home in or on the tree? Describe what you see.
____
____
____

6. Describe the tree's bark.
____
____
____

7. Are the roots visible? ____

8. How far from the trunk do the roots show above ground? ______________

Name ______________________ Date ______________________

F3: Understanding Chlorophyll

Do you know what "chlorophyll" is? It is the green pigment or coloring matter found in green plants and grass. It is located in tiny cells called "chloroplasts" that are visible through a microscope.

Chlorophyll is important to **photosynthesis**, the process by which carbohydrates are formed in the chloroplasts. The plant is able to manufacture its own food and provide energy to its cells. Certain nutrients, air, water, and light are vital to the process of photosynthesis. In this experiment, you will be regulating the amount of light, water, and air supplied to the chloroplasts in plant tissue.

EQUIPMENT:

4 boxes of same shape and size
Scissors
Pencil
Water and watering can

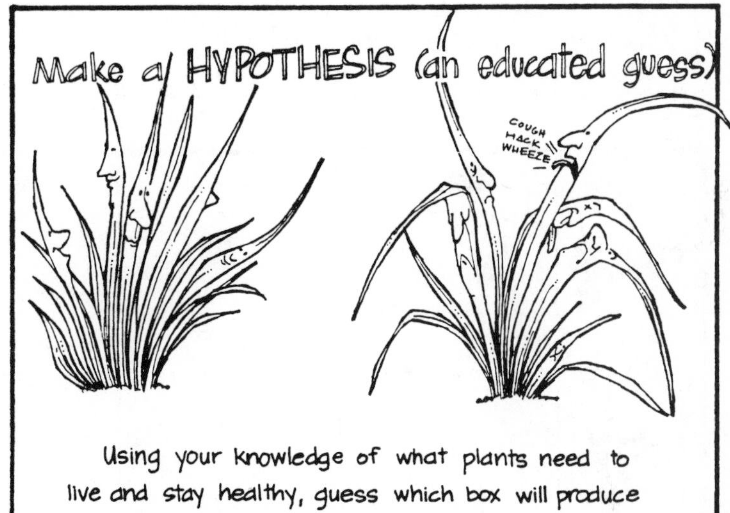

DIRECTIONS:

1. Look at the illustration of the four boxes. Using your pencil and scissors, prepare the lids of your four boxes the same way: Box 1 has 6 holes on top; Box 2 has a circle in the middle of the top; Box 3 has four open squares on the top and one on each of the four sides; Box 4 has no holes. The holes will affect the amount of light and air entering your boxes.

Name ______________________________ Date ______________________

F3: Understanding Chlorophyll (continued)

2. Place all four boxes on top of green grass. Water the grass through the opening in Box 2 only. Do nothing to the rest.
3. Guess which box will produce the healthiest grass. This is your hypothesis.
4. After ten days, lift all four boxes and observe the grass. Record results in the chart. Did the chlorophyll change? Did any grass turn yellow? Which box has the healthiest grass?
5. Replace the four boxes for another ten days. Water the grass in Box 2 only.
6. After ten more days observe the grass beneath the boxes. Record the results in the chart below.
7. Was your "hypothesis" correct?
8. Look up the word "photosynthesis" and record its definition on the back of this sheet.

	BOX 1	BOX 2	BOX 3	BOX 4
RESULTS AFTER 10 DAYS				
RESULTS AFTER 20 DAYS				

Name ______________________________ Date ______________________

F4: Looking for Signs of Photosynthesis

When you looked up the word "photosynthesis," what did you discover? What does a green plant need in order to manufacture or make its own food? Green plants take in carbon dioxide from the air. They need sunlight, water, and nutrients to make food. Then they can successfully release oxygen into the air. Carbon dioxide and water combine to make simple sugar and oxygen. Sunlight is necessary for the plant to form the sugar and for it to release oxygen in this cycle.

In this investigation, you will examine four different types of specimens to determine whether they have chlorophyll and whether they manufacture their own food in the photosynthesis process. After observing these specimens, you will see that the green coloring matter called chlorophyll is not present in all. You will also observe another difference—some contain spores that aid in reproduction, but not in food production.

EQUIPMENT:

Four different specimens: nonpoisonous mushroom, fern, elodea leaf (available in fish/pet shop or order from science supply company listed in appendix), and leaf from a nonpoisonous tree

Hand lens

If sunlight is necessary for photosynthesis a plant that shows signs of photosynthesis (i.e. chlorophyll) probably lives where sunlight is plentiful. However, Elodea is an aquatic plant. What do your findings about Elodea tell you about its natural habitat? Does it thrive in deep or shallow water? What else can you tell about a plant's habitat by simply looking at it?

DIRECTIONS:

1. Use the hand lens to carefully examine all specimens. Look for spores, color, and type of specimen.

Name ______________________________ Date ______________________________

F4: Looking for Signs of Photosynthesis (continued)

2. Record all information on the chart. Write "yes" if you think the specimen manufactures its own food. Do the same for "Presence of Chlorophyll or Spores." Look up "spore" and write the definition on the back of your paper.
3. Think about where you find mushrooms. Are they in the shade or the sun? Are they less likely to need the sun's energy than the other species?
4. In the fall certain trees shed their leaves and chlorophyll is absent from their leaves. In which season would those deciduous trees rest?

SAMPLE	PRESENCE OF CHLOROPHYLL	PRESENCE OF SPORES	MANUFACTURE OWN FOOD
MUSHROOM			
FERN			
LEAF FROM TREE			
ELODEA			

Name ______________________________ Date ______________________

F5: How Moisture from Leaves and Soil Condenses

EQUIPMENT:

6 small glass jars
6 large glass jars
Water
Moist outdoor soil
Assorted fall leaves
1 small nonpoisonous plant

DIRECTIONS:

1. Fill the 6 small glass jars as follows:
   Jar 1: fill with water
   Jar 2: fill with moist outdoor soil
   Jars 3, 4, 5: fill with assorted leaves, separating colors
   Jar 6: fill with green plant and small amount of soil
2. Invert the larger jars, over the filled smaller jars, as shown:

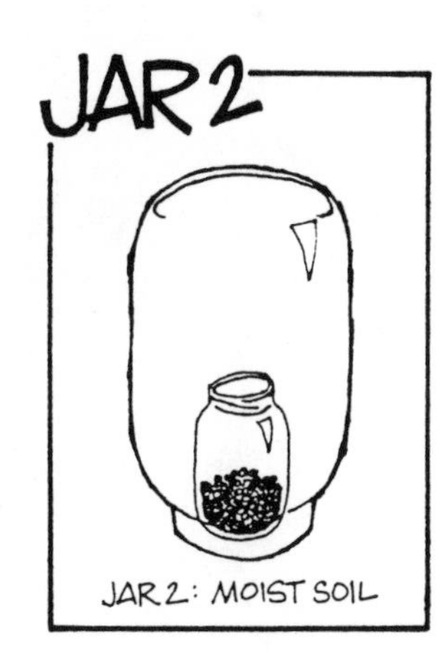

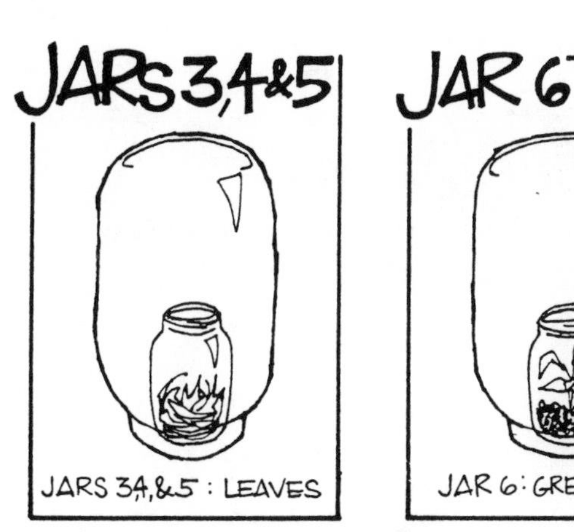

3. Put the jars outdoors in the same location and observe over several days.

Name ________________________ Date ________________________

F5: How Moisture from Leaves and Soil Condenses (continued)

CONCLUSIONS:

1. Describe what happened in each of the six jars. Where did you observe moisture on the jars?
2. Did all 6 jars have the same amount of moisture?
3. Do you think location was important to the amount of moisture found on the jars? You may want to repeat the investigation indoors and see if the results are the same.
4. Think! In the fall, the leaves and soil combine with the moisture (water, precipitation) and assist in the process of decay. As insects and small animals nibble on the leaves, the natural organic compost gets broken down even further. This benefits the soil by helping it hold water longer, thereby enhancing its nutrients, and providing ground cover. Did your jars with soil and leaves show signs of moisture in the form of condensation?

Name ______________________________ Date ________________________

F6: Experiment in Chromotography

In this activity, your class can investigate the effects of chromotography, using leaves to examine chlorophyll and color separations. This should be done as a demonstration, as it involves using a chemical solvent. The solvent can be ordered directly from the supplier listed in the Resource List of this book, or mixed from a solution of 90% petroleum ether and 10% acetone. Be careful when handling or mixing any combustible solvent or in storing and discarding the solution.

In discussing the results of the experiment, lead the children to observe and conclude the following:

Leaves change color because the green pigment matter called "chlorophyll" changes. Some trees, called "deciduous" trees, shed or drop their leaves in the fall. The leaf stem where the leaf is connected to the twig grows a cork-like substance in the fall that blocks the leaf from getting moisture and nutrients from the roots and trunk of the tree.

Due to this blockage, the leaf changes color and starts to die. The warm sunny days, and dryness, and cool nights combine to create the autumn golds, yellows, reds, and browns we see in deciduous trees. The leaves change color because of this blockage and absence and change in the pigment chlorophyll. As the colored leaves in this experiment mix with the solvent, the pigments are released onto the chromotography paper. The colors separate, showing that some leaves contain little or no chlorophyll (as indicated by the color green on the color separation chromotography paper).

EQUIPMENT:

- 3 different colored leaves
- Scissors
- 3 stirrers
- 3 small dishes
- 3 eye droppers
- 3 glass jars with lids
- Liquid solvent, available from Biological Supply Company, see Appendix for ordering information (CAUTION: Use in well-ventilated area)
- 3 strips of filter or chromotography paper

DIRECTIONS:

1. Use the scissors and cut the leaves into small pieces, separating them by color into the glass dishes. Label the dishes.
2. Take the mixed solvent and put a few drops in each of the three glass dishes of cut leaves.

3. Using the three stirrers, stir the contents of each dish separately.
4. One dish at a time, take the eye dropper and suction a few drops of solution from the dish. Slowly release the fluid onto a strip of filter or chromotography paper.

5. Place the three strips of paper into the three glass jars and put the lids on top. Be sure the jars are labeled.

6. Watch and observe the separation of colors found in the solvent from the cut leaves.

CONCLUSIONS:

Describe and discuss the chromotography color separations.

F7: Starting a Leaf Exchange Program

Fall is a great season to begin an exchange program with another school either within your district or outside your region and state. Use the sample letter to introduce your class to a Leaf Exchange Program. The idea is to learn about different fall leaves in your area and other areas of the country at this time of year. Your students will also benefit from a class project and make some new friends in other areas and schools.

The results of the leaf exchange program can be displayed in a bulletin board, as below. Or, the students may want to make Notebook or Scrapbook Collections of different leaves in their area, writing mini-reports about the leaves, taping samples in the book, and including a map of their school area or state.

A SAMPLE BULLETIN BOARD

The sample letter includes a few guidelines about writing a business letter, including reminders to the students about proper punctuation. A letter can be drafted from your class by the students and sent to one or more school districts.

September 20, 1990

Mr. Brown's Class
Northwood School
12 Northwood Drive
Portsmouth, Virginia 23703

THIS IS A BLOCK FORM OF A BUSINESS LETTER
DO NOT INDENT MARGINS!

The Superintendent of Schools
West Hartford Public Schools
West Hartford, Connecticut 06117

A COLON (:) IS USED HERE

Dear Sir:

Our class is studying about the leaves and trees in the climatic seasons. We want to participate in a Leaf-Exchange Program with students in your school district.

If you could help us find a class in your school system, we would appreciate your help very much.

In exchange for the leaves and projects your school system's class will make, we'll send a sample of leaves from our area in a form they will find enjoyable and educational.

Thank you kindly for your support and help.

DON'T FORGET THE COMMA IN THE "CLOSING"

Sincerely,
Johnny Richards and Classmates
Grade 5

Name ____________________ Date ____________________

F8: Observing Precipitation in the Fall

The cooler nights and impending frost of fall show signs of precipitation in our yards. Think about the morning dew and how visible it is on the grass and leaves. Think about the weather and the changes in temperature upon various forms of precipitation from the clouds. How does the change in temperature affect the forms of precipitation from the clouds? How can the cooler temperatures affect rain?

In this investigation, think about the same results of the water molecules in water and ice and the changes in matter that occur. Think about what effect salt has on the freezing point of water.

EQUIPMENT:

1 empty metal juice can
Ice
Stirrer
Water
Salt

DIRECTIONS:

1. Fill the can three-quarters full of ice.
2. Pour water over the ice to fill the can close to the top.
3. Add salt and stir.
4. Observe what happens to the water, ice, and can.

CONCLUSIONS:

Write a paragraph describing the changes in matter of the ice and water. What appeared on the can? What do you think the salt did? You may want to try the experiment again without using any salt.

F9: Going on a Fall Scavenger Hunt

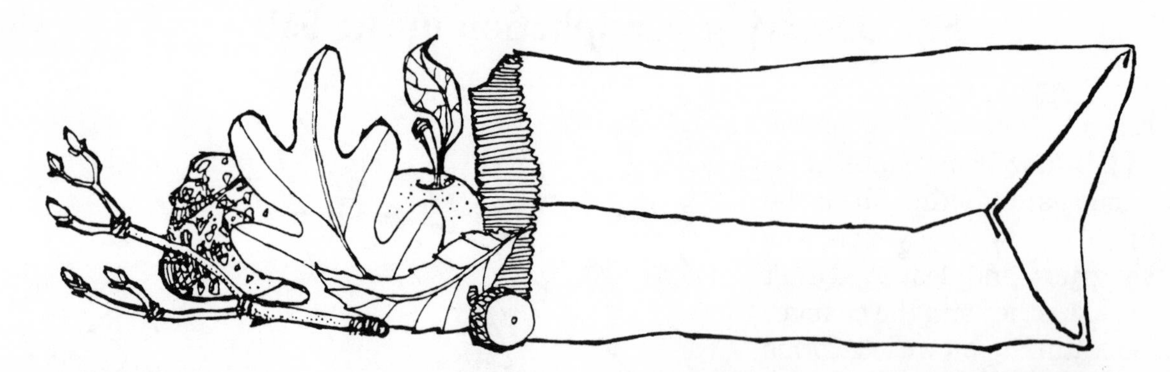

Each season the students will have the opportunity to go on a scavenger hunt. The lists are different for each season. Some items enumerated on the list are to be located or identified only and not picked. Discuss the items with the class before going on the hunt.

The hunt may be done in school or at home. It may be done on an individual or small group or team basis. Teams should have five or six students and a captain.

The captain of the team stays in a central location where the team members will return when they find items on the list. The captain should verify the specimens of collection so the team members do not duplicate their efforts.

Before beginning, the class should add a few items to the list of items indigenous to your region or area in the fall season. Try to add four items to the list for the students to locate outside.

FALL SCAVENGER HUNT LIST

Name______________________

Team Captain (optional)______________________

Members on My Team______________________

Four Extra Items______________________

FALL SCAVENGER HUNT LIST

1. ACORN
2. LEAF (Evergreen or Deciduous, identify which)
3. INSECT*
4. LEAF (with opposite vein arrangement)
5. STEM WITH TERMINAL BUD*
6. SIMPLE LEAF (each leaf stem has one blade)
7. COMPOUND LEAF (each leaf stem has several blades)
8. MONARCH BUTTERFLY*
9. LEAF (with chlorophyll)
10. MIGRATING BIRD*

*LOCATE, BUT DON'T PICK OR TAKE

Name ______________________ Date ______________________

F10: The Harvest Moon and the Equinox

The harvest moon is a full moon that generally occurs near the Fall Equinox, around September 23. During the Equinox, the moon, earth, and sun are in a direct line.

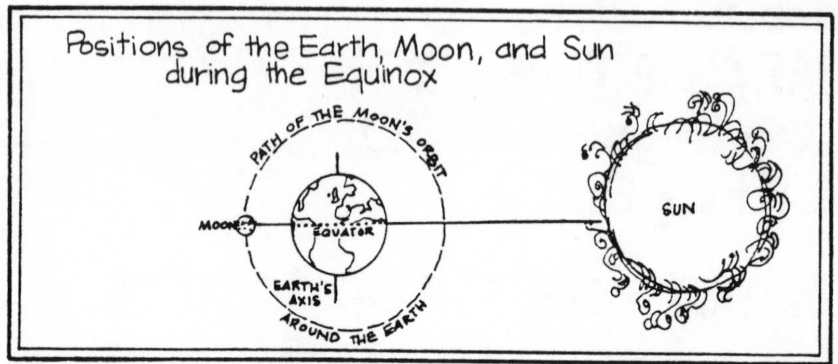

For this activity, you will complete the moon observation chart. Look at the illustration below to help you identify the phase you are observing. Go outside at night with a parent and observe the moon for a few moments. Record other important information on the chart, too, such as the weather conditions. Also read your local newspaper to verify the exact time of moonrise in your area. NOTE: Be sure you have parental supervision when outdoors.

EQUIPMENT:

Copy of Moon Observation Chart, Pencil, Newspaper

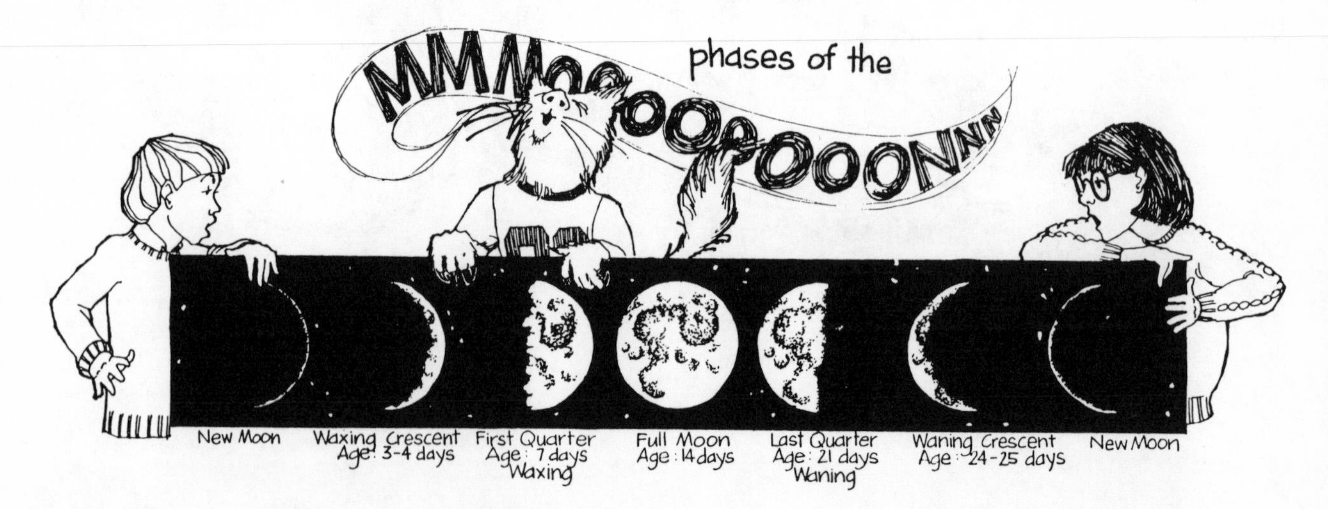

Date	Forecast	Time of Moonrise	Phase of Moon
1.			
2.			
3.			
4.			
5.			
6.			
7.			

Date	Observation	Actual Weather at time of observation
1.		
2.		
3.		
4.		
5.		
6.		
7.		

Name ______________________________ Date ______________________________

F11: Examining Fall Weather

Using the "Seasonal Diary of Weather Statistics," observe and record the weather in your area during the fall. Complete your observations for one week, observing and recording fluctuations in temperature, the precipitation, and any other significant weather events.

You might want to include a copy of your weather findings in an exchange package with students from other school districts in different parts of the country, as with the Leaf Exchange Program.

EQUIPMENT:

Copy of the Seasonal Diary of Weather Statistics

Pencil

Name ______________________

A SEASONAL DIARY OF WEATHER STATISTICS

Season : ______________________
Month : ______________________
City : ______________________ State : ______________________
School: ______________________
Grade: ______________________
Teacher: ______________________

KEY: Cloudy — Snow — Rain — Sunny

Dates	Temperature High	Temperature Low	Difference	Precipitation (See Key)

Weekly summary ending ______________________ (date)

Other significant findings or weather phenomenon (i.e., a severe storm or hurricane) describe on back of this sheet.

Name ______________________ Date ______________________

F12: The Fall Sky—Looking for Pegasus

Look up "Pegasus" in the dictionary and write its definition on the back of this sheet.

In this activity, you will make a Constellation Viewer using either an oatmeal box or a styrofoam paper cup, depending on the size viewer you want.

EQUIPMENT:

- Empty oatmeal box (18-ounce cylinder) or styrofoam cup
- Flashlight
- Sharp pencil (CAUTION: Use with care)
- Patterns for Pegasus
- Scissors
- Tape

DIRECTIONS:

1. Select the box and the appropriate pattern. Cut out the paper pattern and place it on one end of your cup or box. Tape it in place.
2. Using the sharp pencil, carefully punch holes the SAME SIZE as the pattern shows. Keep the point of the pencil down and away from your eyes.
3. Take the cup or box to a totally darkened room. Turn on your flashlight and aim it at an angle toward the top of the box or cup. Light will penetrate the openings and the pattern will appear on the ceiling or wall. You can adjust the focus by moving closer. Look for the constellation in the evening sky, too.

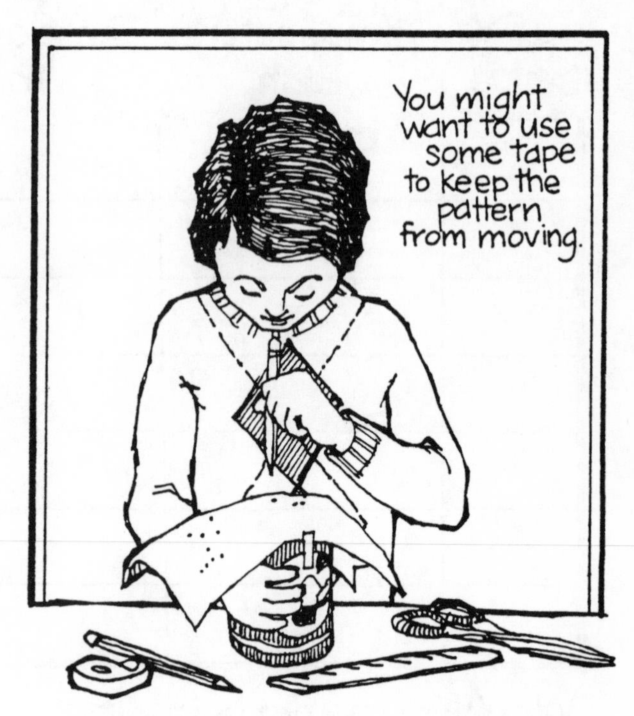

PATTERNS FOR PEGASUS

Use this pattern if you are using a styrofoam cup:

Use this pattern if you are using an oatmeal box:

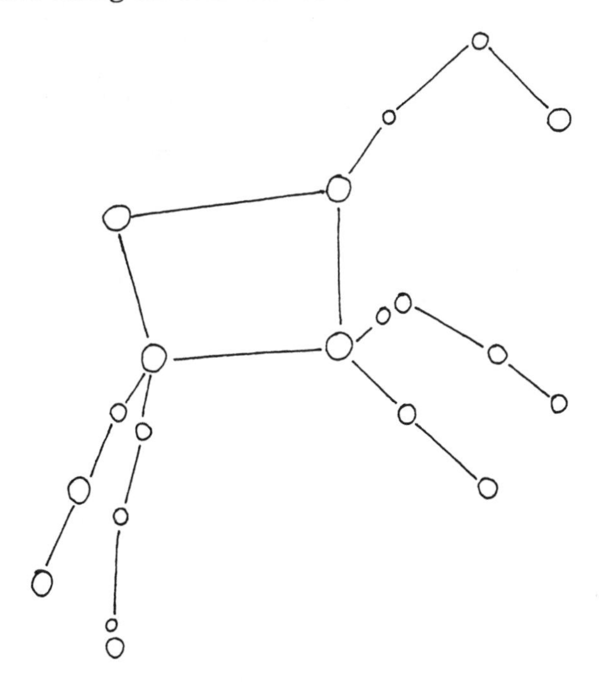

FALL
enrichment
activities

Name ______________________ Date ______________________

Fall Cookery—Solar Brewed Decaffeinated Tea

EQUIPMENT:

- $^1/_2$-gallon glass jar with screw cap
- Decaffeinated tea bags
- Water
- Lemon
- Honey or sugar
- Cup or mug
- Spoon
- Sunshine

(Special thanks to Jonathan Craig, Director of Ecology at Talcott Mountain Science Center in Avon, CT for this recipe!)

DIRECTIONS:

1. Fill the jar with water.
2. Place two or three tea bags in the water.
3. Screw the lid on tightly.
4. Put the jar in a warm sunny spot, like a windowsill or outdoors, where there is a lot of sun.
5. How long does it take to brew the tea?
6. Serve with lemon and/or honey or sugar.

THINK:

What is the role of the autumn sun in brewing this tea? Do you think the results would be the same in other seasons? Try it and see!

Name ______________________________ Date ______________________

Pumpkin Seed Snack

EQUIPMENT:

Small pumpkin with top removed
Salt
Water
Parmesan cheese (optional)
Butter
Heat source (CAUTION: Use only under adult supervision)
Foil-lined baking tray
Hot mitts
Saucepan
Spoon
Preheated 300° oven (CAUTION: Use only under adult supervision)

DIRECTIONS:

1. Scoop out the pumpkin seeds with the spoon. Wash seeds.
2. Soak seeds overnight in a solution of 2 teaspoons of salt and 1 cup of water. What happens to the seeds as they absorb water?
3. Pat seeds dry next day. Melt 1/2 cup of butter in a pan. Sprinkle Parmesan cheese in pan. Put in seeds and stir to coat.
4. Spread coated seeds on foil-lined tray. Bake in oven 1/2 hour.

THINK:

Why do you soak the seeds overnight? When are pumpkins harvested? How long is their growing season? Pumpkin seeds are nutritious and a good source of protein and phosphorus. Find out how protein and phosphorus help your body.

Did you know that seeds have *dormant* or sleeping stages in their cycle? Try planting a few pumpkin seeds directly from the pumpkin in the soil. As the seed germinates, water absorbed through the coat helps it to swell and grow.

Name ______________________ Date ______________________

Leaf Rubbings and Printing

EQUIPMENT:

- Assorted leaves
- Tempera or poster paints
- Brushes
- Paper towels
- Paper
- Crayons and chalk (optional)

DIRECTIONS:

1. Examine your leaves for the patterns of veins on them.
2. Paint the underside with paints and brush. Coat well.
3. Turn the leaf over onto the paper and press gently for a few seconds. Lift the leaf directly up, carefully.
4. Repeat process with other leaves to create a design or pattern.
5. Try another medium like chalk or crayons to make a rubbing. Place the leaf, vein-side up, beneath paper of lighter texture.
6. Take the crayon or chalk and rub with the side of the chalk or crayon. The impression of the leaf should be visible on the paper.

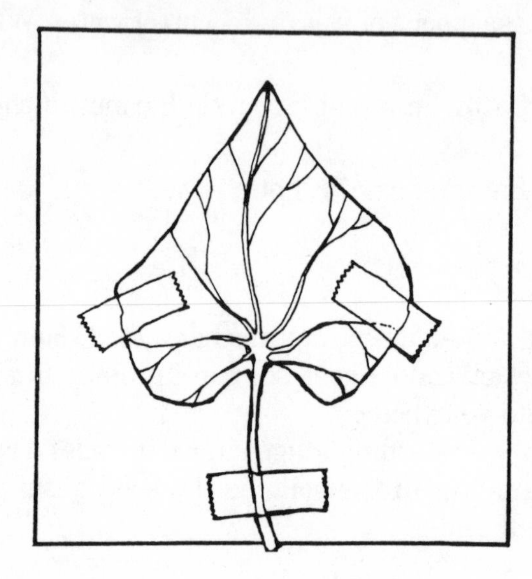

To keep the leaf from moving while you make your rubbing, you can anchor the leaf to the paper (vein side up!) with some scotch tape.

THINK:

How are leaves different in the fall in your area? In other parts of the country? What are the names of the trees that shed their leaves in the fall? Why do leaves change color and what happens to the tree during the winter?

Name________________________ Date________________________

FALL SCIENCE STORIES, SONGS, & POETRY

– MAKING A FAMILY TREE –

One way to study any living thing is to make a family tree for it and see how the different types of that plant or animal are similar and how they differ. Here's an example:

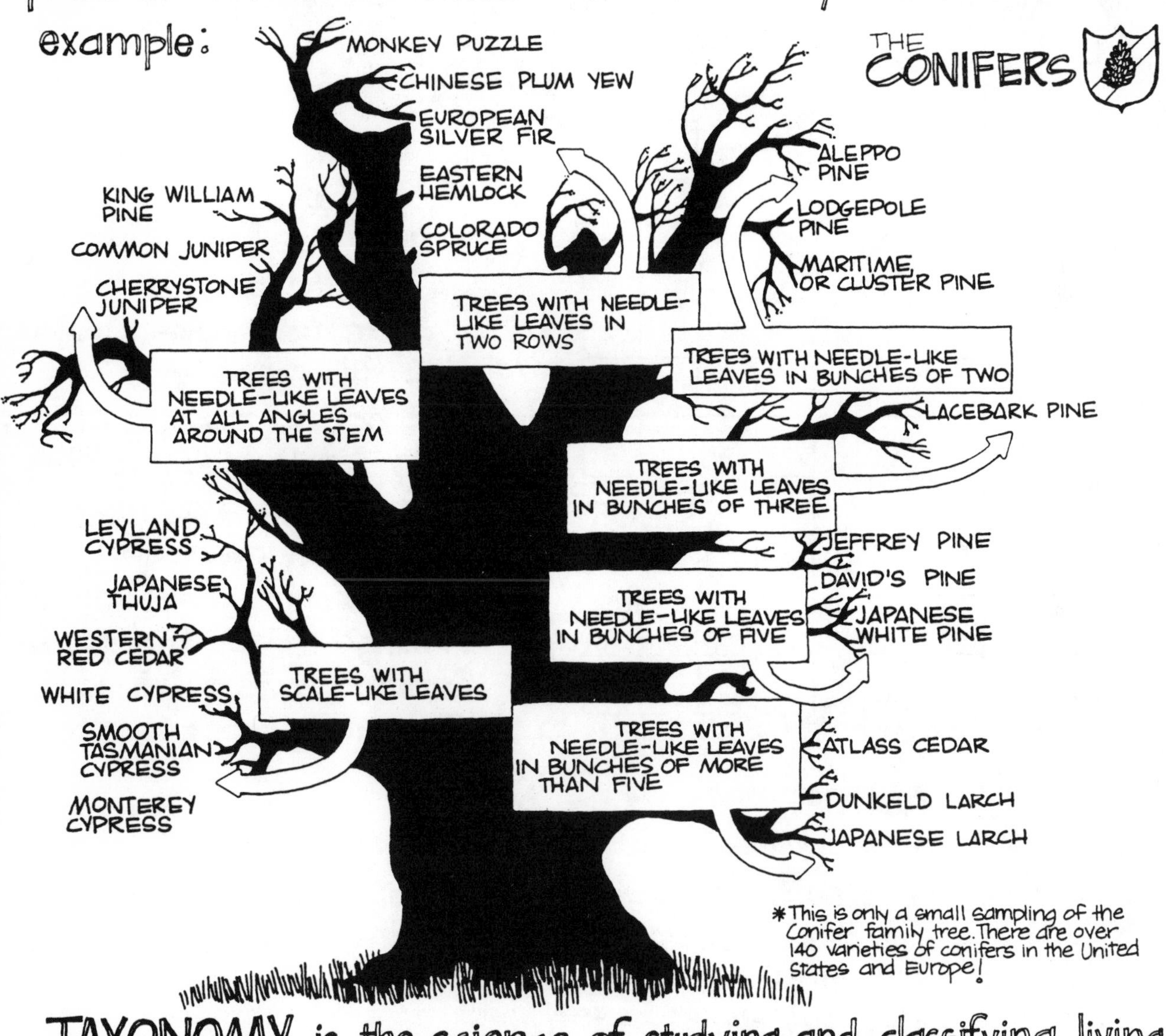

*This is only a small sampling of the Conifer family tree. There are over 140 varieties of conifers in the United States and Europe!

TAXONOMY is the science of studying and classifying living things. How are these trees grouped? What does this tell us about how taxonomists classify the animals and plants they study? Try making a family tree for Broadleaf trees.

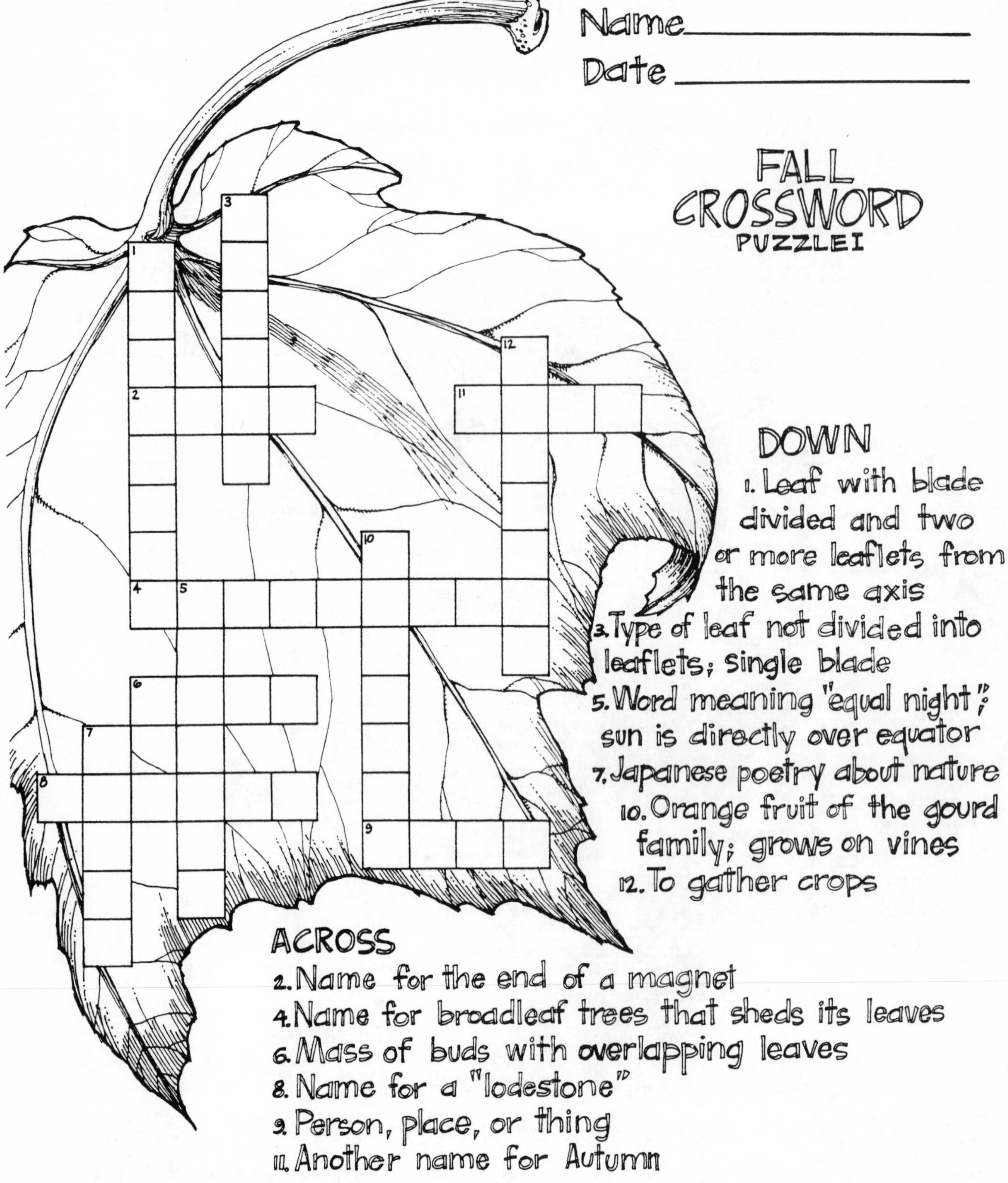

Name______________________

Date ______________________

FALL CROSSWORD PUZZLE I

DOWN

1. Leaf with blade divided and two or more leaflets from the same axis
3. Type of leaf not divided into leaflets; single blade
5. Word meaning "equal night"; sun is directly over equator
7. Japanese poetry about nature
10. Orange fruit of the gourd family; grows on vines
12. To gather crops

ACROSS

2. Name for the end of a magnet
4. Name for broadleaf trees that sheds its leaves
6. Mass of buds with overlapping leaves
8. Name for a "lodestone"
9. Person, place, or thing
11. Another name for Autumn

Name ______________________ Date ______________________

Different Leaves: Unscramble Me!

These four descriptions of deciduous tree leaves are to be matched with the leaf that describes each one. First, unscramble the letters to find the answer that goes in the blank. Then look at the four mixed-up leaf drawings and draw a line from that leaf to the correct spelling of the leaf you wrote on the blank.

DESCRIPTION:

1. My leaves grow opposite on the trees. One grows this way; one grows that way! My buds are red all winter; my leaves are red and yellow in the fall. I'm often tapped for something yummy on your pancakes! *Unscramble:* APEML

   I am a ______________________ tree.

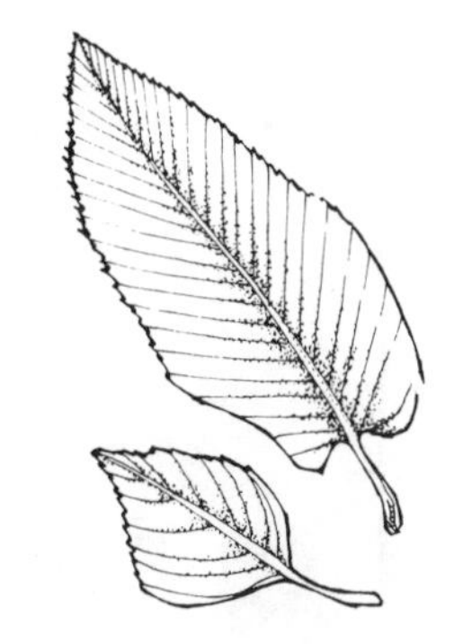

2. I grow in clusters. I like sunny dry places. My leaves turn golden in the fall, and beavers love to nibble on my bark. They even use my twigs in their homes. *Unscramble:* NEPSA

   I am an ______________________ tree.

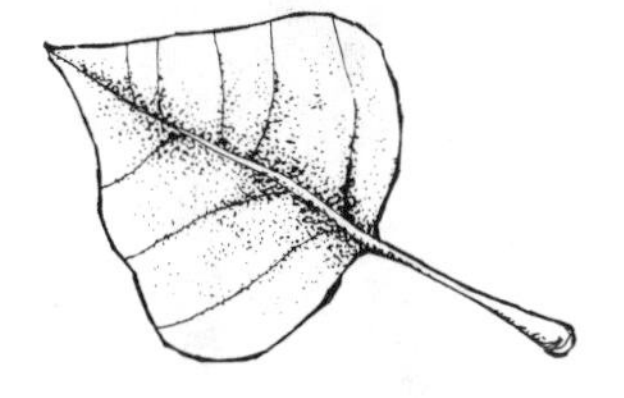

3. I am famous for the contrasts on my trunk. Artists love me in their fall paintings. People show me off in their fireplaces, but don't generally burn me. My leaves are yellow in the fall, and I shed them. *Unscramble:* IHCRB

   I am a ______________________ tree.

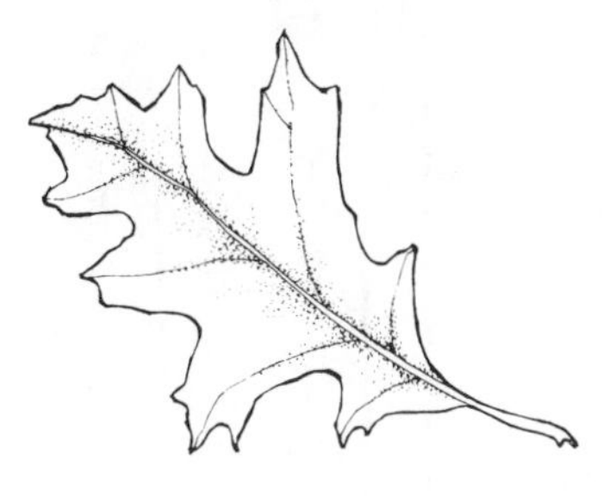

4. My leaves turn dark brown or red in the fall. I have acorns and am known for my strength. My leaves are simple and alternate. *Unscramble:* AKO

   I am an ______________________ tree.

Name ______________________________ Date ______________________________

WRITING FALL HAIKU

"Haiku" is a type of poem from Japan. It refers to the changes in nature during the seasons of the year. The verse does not have to rhyme. Here is a simple haiku that follows the pattern of five-seven-five syllables. There are five syllables in the first line, seven in the next, and five in the last.

LEAVES
SEE THE CHANGING LEAVES—
GREEN, GOLD, RED AND BROWN, I SEE—
BLOWING IN THE WIND.

On the back of this sheet, write your own haiku about the fall season. Some suggestions are: Harvest Moon, Equinox, Fall Weather, Changes in Insects and Animals.

Name________________________ Date____________________

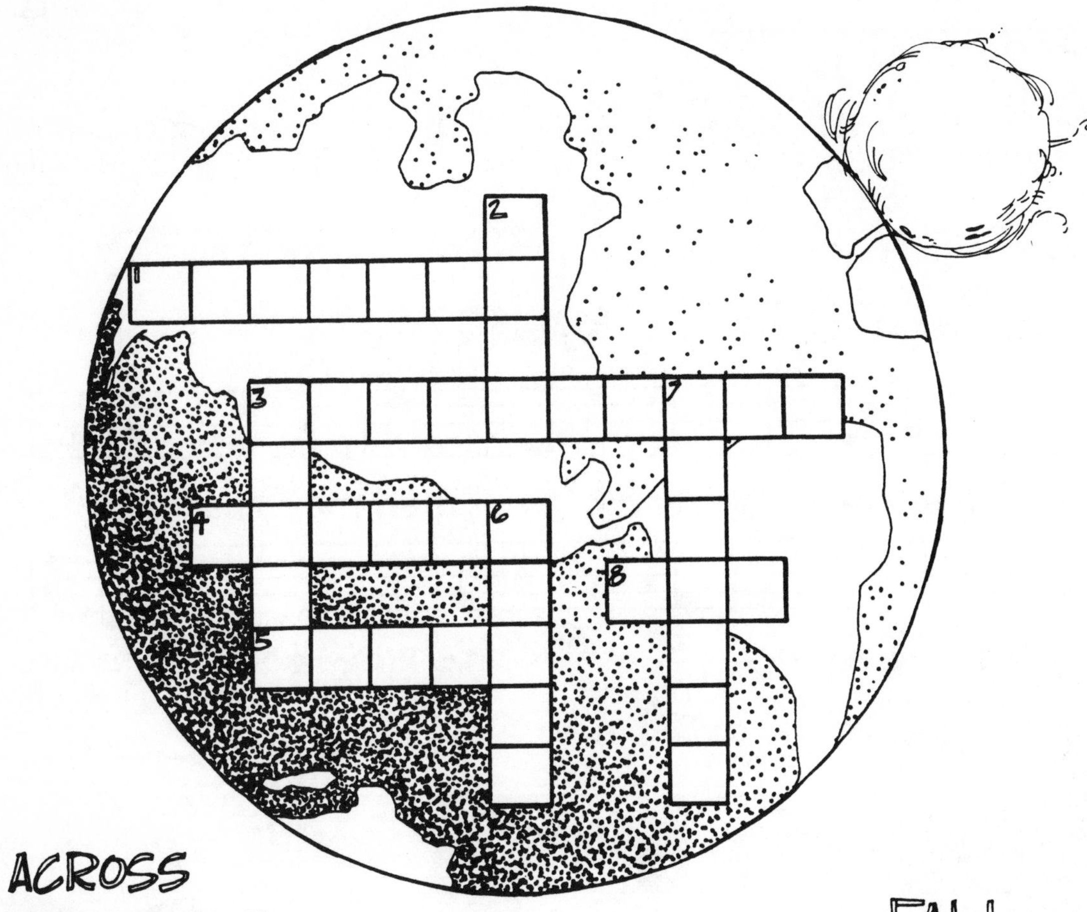

FALL CROSSWORD PUZZLE II

ACROSS

1. Time when the sun crosses the Equator
3. Half of the earth
4. A season
5. The star 93,000,000 miles from earth
8. A period of lightness

DOWN

2. Imaginary line around which the earth rotates
3. A measure of time (plural noun)
6. A time of darkness
7. Imaginary circle dividing the earth into two hemispheres

Music by: Patricia McCamish Donohoe Lyrics by: Julia Moutran

The Winter Season

TEACHER'S GUIDE FOR WINTER

The second section of this book contains activities for the winter season. If you do not live in a cold climate, you can still do all of the activities with your students, excluding activity W2.

As in the first section, there are twelve major lessons or activities (W1 through W12). These are followed by nine enrichment activities, including making pocket sundials, solar food warmers, a winter song, and a crossword puzzle, as well as opportunities to apply science concepts in other curriculum areas, including language arts.

Be sure to use the Student Progress Flow Chart so your students can keep a record of their progress. You might also want to send a copy of the progress chart home with your report cards or share them with parents during conferences.

There are several activities in Section Two that are similar to activities in Section One. These include examining winter trees, looking for another constellation (Orion), going on a winter scavenger hunt, and finding out about the Winter Solstice. There is also the opportunity again to make a science notebook. Reproduce the full-page illustration on page 42 as the notebook's cover, and use it to discuss the following concepts with your students:

- Rabbits are nestling to keep warm.
- Trees and leaves have changed since fall.
- Frog has burrowed in mud for winter.
- Groundhog is hibernating.
- Sun is setting earlier—shadows are longer.
- The brook is starting to freeze.
- Animal tracks appear in the snow.
- Children are hanging bird feeders—food harder for birds to find.
- Bulbs are dormant.
- Low-lying stratus clouds are typical of winter clouds and sky.
- Children's clothing indicates temperature change.

For your audio-visual needs, check the Coronet/MTI list of films and science videos recommended for these activities. The list is found in Appendix 3. You may also want to order the book, *The Story of Punxsutawney Phil, the Fearless Forecaster*, also listed in the Appendix. This is a good time of the year to discuss animals—their preparations for survival during the winter months and how their bodies adjust to the change in temperature. Groundhogs are unique in their hibernation and adaptation to their environment and the cold weather.

Send a copy of the *Parent Letter* to your students' parents, explaining the types of activities in this unit for study and enrichment.

Use a variety of learning settings with your student. Talk about the changes in their areas during each different season. What activities can they do during the winter months that are different from those they did during the fall? Are certain clothes, sports, foods, entertainments, and vacations different at this time of the year?

Compare the findings of the "Winter Scavenger Hunt," "Examining Trees in Winter," and "The Winter Sky" with the corresponding fall activities. Are there similarities? Are there differences? Lead your children to make these important comparisons and observations about their world.

Have your students look for newspaper articles about the Winter Solstice, Groundhog Day, and Winter Weather. This will reinforce their learning. Make a special bulletin board to display these articles and their own science lab or activity sheets.

Decorate your classroom or a showcase with the beautiful constellation viewers, animal track molds, or pocket sundials. Feature a display of the animal track molds and have a contest for other students to guess the animal tracks. You could have pictures of the animals in the background and a matching contest to guess which print belongs to which animal. Display the children's stories from the activity "Animal Tracks: Who Done It?" These fables and mystery stories can be enjoyed by all ages, as can the song on tracking.

Dear Parents,

As a continued part of our science program, your child is learning about the Winter Season. During this time, your child will learn about:

Changes in the winter trees and leaves;

Changes in the weather;

Observation of wildlife;

Continued observation of the sky and the constellation Orion;

Changes in temperature.

In addition to the above, there will be many activities for science enrichment, including making a pocket sundial and a mini-solar food warmer!

At the end of our unit of study, your child will receive a "Winter Science Season Certificate."

Thank you very much for your interest in your child's science program.

Sincerely,

____

Teacher

W1: What and When Is Winter?

Refer to the first fall lesson, *What and When Is Fall?* Repeat this activity with your students and discuss the position of the earth during the winter. Review these vocabulary words: ellipse, axis, and illuminate.

Be sure to review or discuss these concepts with your students after completing the activity. It is helpful to repeat the experience with them to reinforce the concepts.

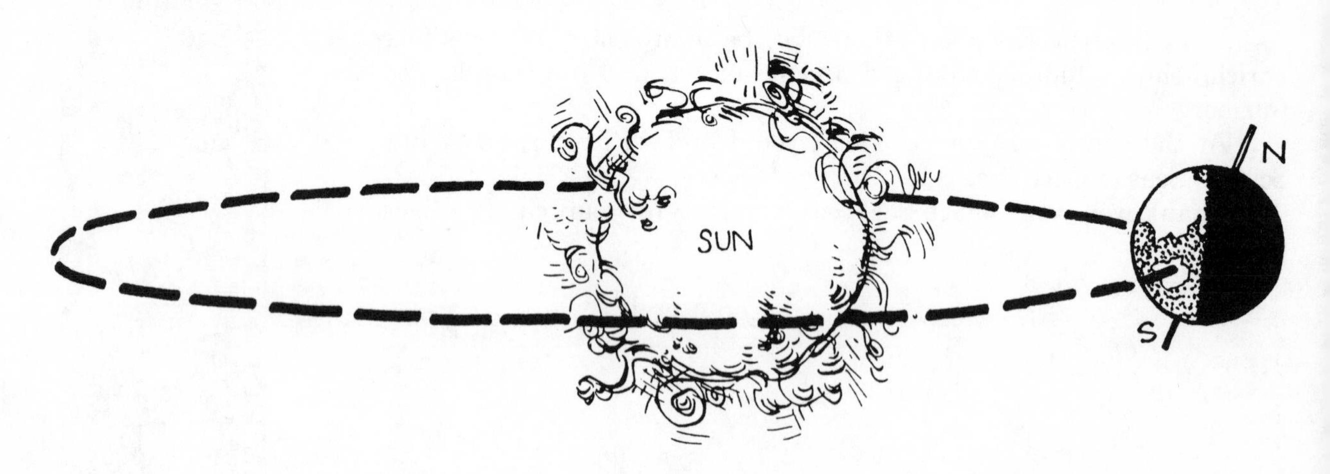

Share these facts with your students after they have completed the activity:

During winter the North Pole has darkness.

The Northern Hemisphere has winter while the Southern Hemisphere has summer.

During winter the sun is lower in the sky and days are shorter; nights are longer.

On the Winter Solstice (around December 22), the sun is lowest in the sky for the Northern Hemisphere.

During winter the North Pole is tilted away from the sun.

On the first of January, the earth and sun are closest in distance.

As a follow-up have your students write about their area during the winter and a contrasting area during the same time of the year. Have them report on differences in climate, trees, weather, and lifestyles. Why do people vacation in certain areas during winter?

Name ______________________________ Date ______________________

W2: Examining the Changing Forms of Matter

Think about the composition of water and how it changes during the winter. Water can be in a liquid form, a solid form like ice or snow, or a gaseous form like steam or water vapor.

When something is in a liquid state, it takes the shape of the part of the container it occupies. The molecules have greater freedom of movement than in a solid form, where the molecules are closer together. In the gas form, the molecules have freedom to move and separate.

In a snowflake crystal, as in many crystals, the atoms are very close together and arranged in a pattern.

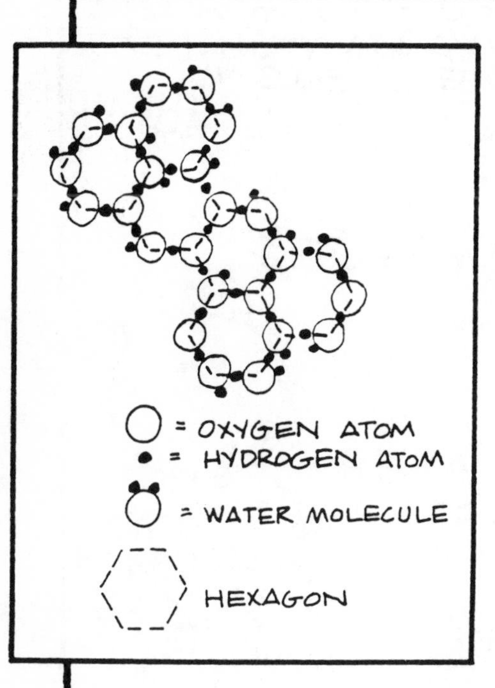

WHEN WATER IS IN ITS SOLID FORM THE MOLECULES ARE "LOCKED" INTO A SPECIFIC PATTERN. HERE IS A DRAWING OF HOW WATER MOLECULES BIND TOGETHER. NOTICE HOW IT LOOKS LIKE A HONEYCOMB MADE UP OF MANY SIX SIDED SHAPES (HEXAGONS).

BEFORE IT MELTS, LOOK AT ONE OF THE SNOWFLAKES YOU'VE CAUGHT. IF IT IS A VERY COLD DAY AND THE SNOWFLAKES ARE NOT STICKING TOGETHER TOO MUCH, YOU MIGHT BE ABLE TO SEE THAT EACH SNOW-FLAKE HAS SIX PERFECT POINTS OR RAYS. IN FACT ALL SNOWFLAKES HAVE SIX SIDES BECAUSE WATER MOLECULES ARRANGE THEMSELVES IN GROUPS OF SIX WHEN WATER CHANGES FROM LIQUID TO SOLID.

Name ________________________ Date ________________________

W2: Examining the Changing Forms of Matter (continued)

In this investigation, you will examine snowflakes as they change their structure and form of matter. In winter, snowflakes and their crystals are formed as the result of changes in the temperature outside. Using different snowflake catchers, see if you can examine the shape of the crystals. Do you see patterns in the flakes and then changes in the form of the crystals as they come in contact with the different materials of the catchers? Do some catchers keep the snowflakes colder and prolong their melting time? Which catcher do you think will keep the snowflakes from melting so fast?

EQUIPMENT:

- Plastic lid top
- Dark-colored felt square
- Dark-colored 5″ × 7″ or larger sheet of construction paper
- Empty tin can with lid removed
- Glass jar
- Magnifying lens
- Stopwatch
- Pencil

DIRECTIONS:

1. You will be examining the melting time of five different snowflake catchers: the plastic lid, the felt square, the glass jar, the tin can, and the construction paper.
2. Go outside on a snowy day with your teacher. Catch a snowflake with one catcher at a time. Use a stopwatch to record how long the snowflake takes to melt on the catcher. Record the number of seconds on the chart shown here by the name of the catcher.
3. Use your magnifying lens to observe the patterns of the snowflake.

WATER IS MADE UP OF VERY SMALL **MOLECULES**. EACH WATER MOLECULE IS MADE OF THREE **ATOMS**. TWO OF THESE ATOMS ARE **HYDROGEN** ATOMS, THE THIRD ONE IS AN **OXYGEN** ATOM.

HAVE YOU EVER HEARD ANYONE CALL WATER "H_2O"? (H_2O IS READ "H-TWO-O.") THE H STANDS FOR HYDROGEN; THE 2 TELLS YOU THERE ARE TWO OF THEM; AND THE O STANDS FOR OXYGEN.

THIS IS A MODEL OF A WATER MOLECULE

H = HYDROGEN
O = OXYGEN

H H O

REMEMBER — IT DOESN'T MATTER WHETHER WATER IS IN THE FORM OF A GAS, A LIQUID, OR A SOLID, ITS MOLECULES WILL ALWAYS BE MADE OF TWO HYDROGEN ATOMS AND ONE OXYGEN ATOM.

Name ______________________________ Date ______________________

W2: Examining the Changing Forms of Matter (continued)

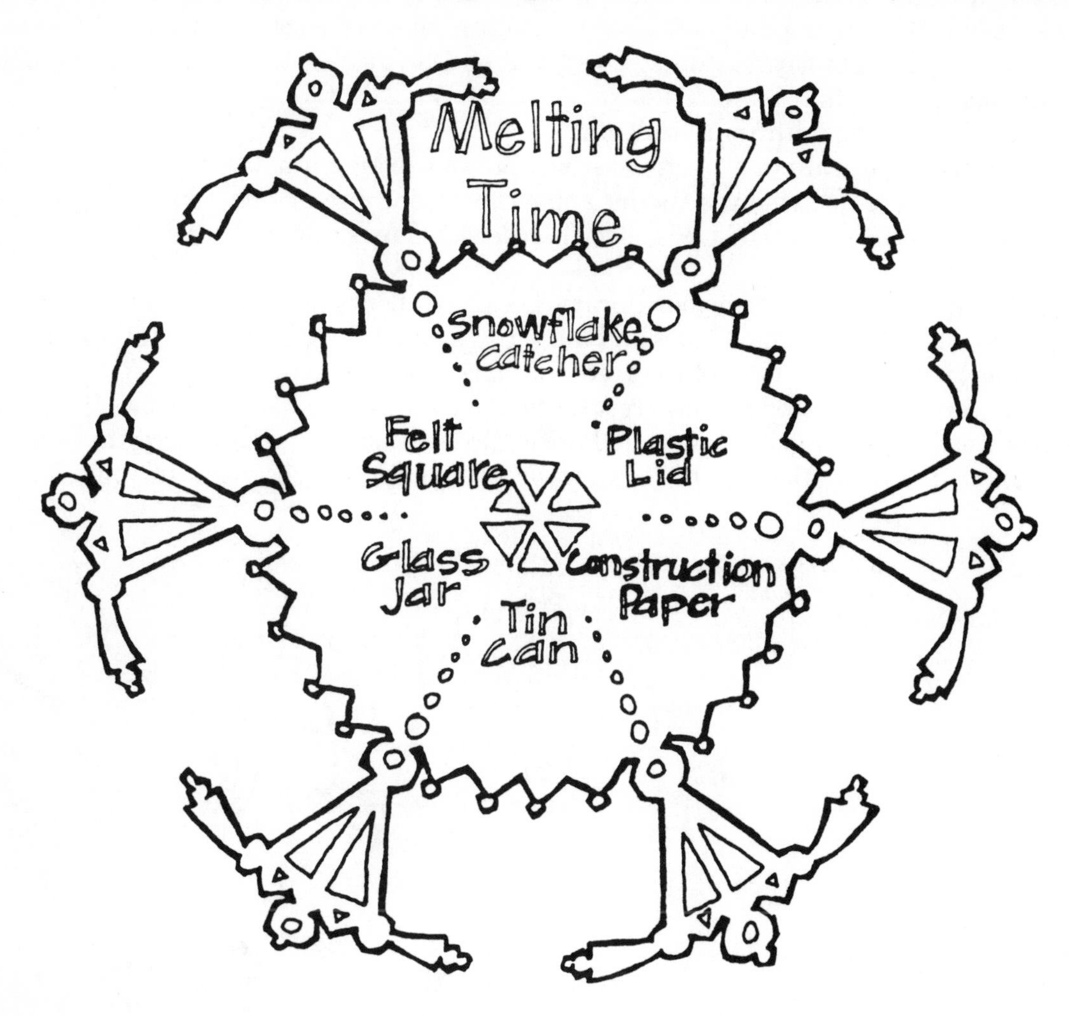

CONCLUSIONS:

1. Did the type of catcher you used make any difference in the melting rate of the snowflakes as measured by the stopwatch?
2. Did the size of the snowflake make any difference in the melting rates?
3. The snowflake changed from a ____________________ to a ____________________.

W3: Examining the Trees in Winter

If you took your class on a fall tree observation walk, here is an excellent opportunity to compare the differences in trees during different seasons. Discuss these concepts with your students after going on a winter tree observation walk. Use "A Tree Observation Chart" for your students to complete during their walk.

Discuss the following with your students:

In winter many trees are in a "resting" stage. Similar to an animal that hibernates, the tree is also dormant. There are changes in the appearances and production levels of trees during different seasons. In the fall, many trees shed their leaves. Trees that shed their leaves are called "deciduous" trees.

Get a field guide on trees from your library. Let the students read the guide to see how trees are categorized. Show them maps of trees and teach them how to look at trees. Winter provides the perfect setting to observe the form of the tree—its bark, its twigs, its height.

Using a map of the United States forest belts, have the class make a plaster of paris raised relief map of the different major forests: the Pacific, Rocky Mountain, Plains, Northern, Central, and Southern. This information can be found in pocket guides to trees found in your library.

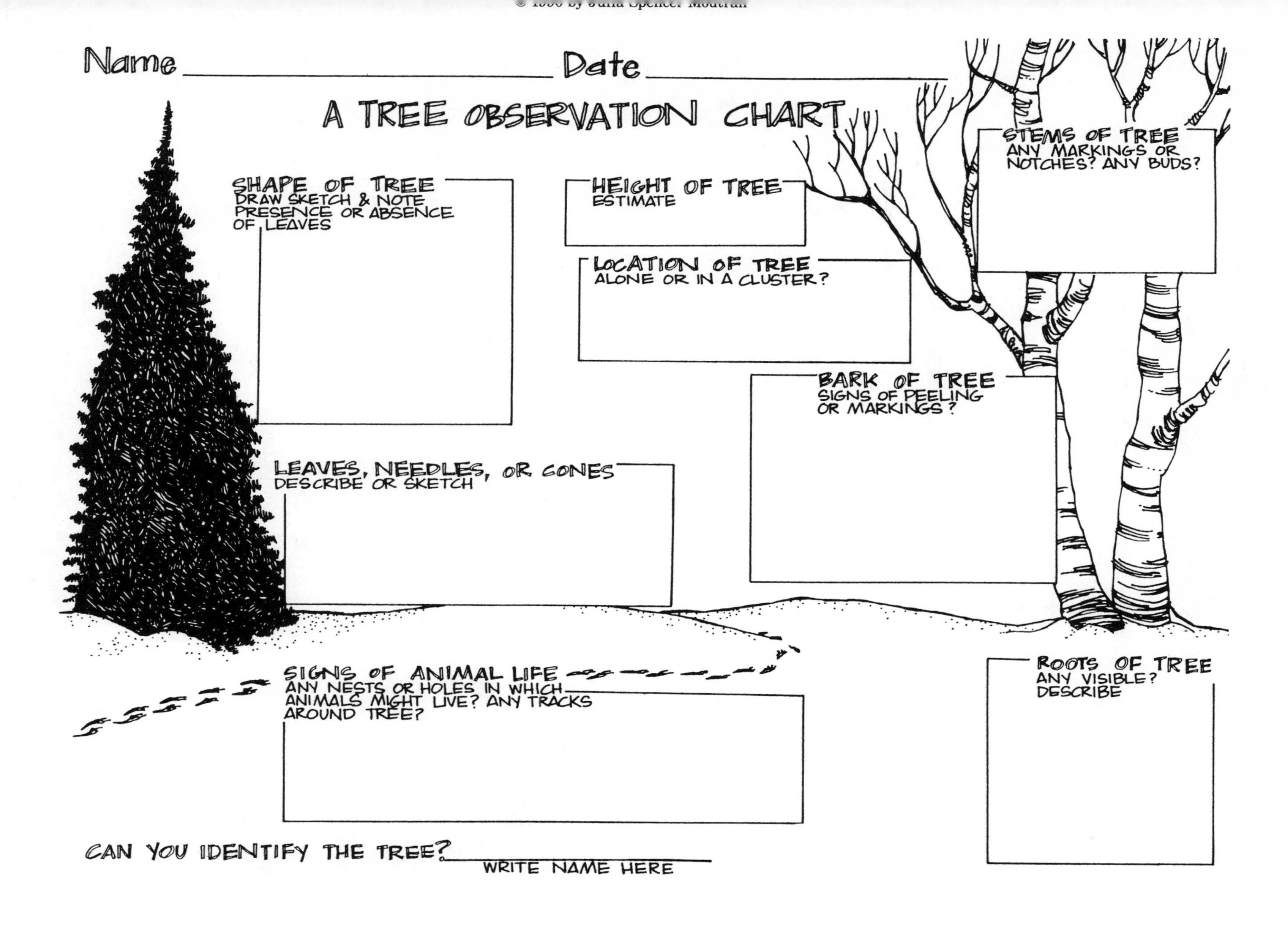
Name
Date
A TREE OBSERVATION CHART
SHAPE OF TREE
DRAW SKETCH & NOTE PRESENCE OR ABSENCE OF LEAVES
HEIGHT OF TREE
ESTIMATE
LOCATION OF TREE
ALONE OR IN A CLUSTER?
STEMS OF TREE
ANY MARKINGS OR NOTCHES? ANY BUDS?
BARK OF TREE
SIGNS OF PEELING OR MARKINGS?
LEAVES, NEEDLES, OR CONES
DESCRIBE OR SKETCH
SIGNS OF ANIMAL LIFE
ANY NESTS OR HOLES IN WHICH ANIMALS MIGHT LIVE? ANY TRACKS AROUND TREE?
ROOTS OF TREE
ANY VISIBLE? DESCRIBE
CAN YOU IDENTIFY THE TREE?
WRITE NAME HERE

Name ______________________________ Date ________________________

W4: Observing Changes in Temperature

The many forms of precipitation we observe during winter are caused by varying changes in the temperature of the atmosphere. As the temperature drops, the chance of rain changing to sleet or snow increases.

In areas of the country where snow and ice accumulate on the walkways and roads, salt is applied to melt the ice. The freezing point of water is 32°F. When you add salt to water, the solid form of ice or snow changes to a liquid. In this investigation, you will observe the effect of salt on ice. Salt actually lowers the freezing point of water, making the ice melt.

EQUIPMENT:

- 1 tray of ice cubes
- 2 (2-cup size) glass measuring cups
- 2 thermometers
- Box of salt
- Ice crusher
- Spoon

DIRECTIONS:

1. Crush the ice cubes into small pieces.
2. Fill both glass cups with the same amount of crushed ice.
3. Add ½ cup of salt to one of the cups of ice. Mix with the spoon.
4. Insert the two thermometers and take the readings. Record.
5. Wait several minutes, then observe and record the temperature.

CONCLUSIONS:

What did the salt do to the ice? How about the temperature? Try this again without crushing the ice. Will it react more quickly on smaller pieces of ice?

W5: Freezing and Melting

No matter where you live, winter is a time to think about changes in forms of matter as a result of changes in temperature.

When we think of winter in the colder areas of our country, we think of the obvious signs of colder temperatures and cold weather. We might visualize the effects of snow, ice, and freezing rain. We might think of the ground outdoors. Frost heaves show how the ground and its pavements have been uplifted by the freezing of moist soil and the expansion of water, when frozen.

In this investigation, your students can observe the effects of temperature, heat, and sunlight on matter. Be sure to discuss the roles of molecules and energy, as well as the changes in weather and temperature on precipitation like snow, ice, and freezing rain. Remind students that changes in matter (solid form of water to liquid form) result in changes in the energy of molecules of water. As ice melts, the molecules speed up and the energy increases.

EQUIPMENT:

- 5 clear plastic cups
- 5 insulated paper cups
- 10 thermometers
- 20 ice cubes
- Clock or watch
- 1 glass measuring cup
- Copies of "Freezing and Melting Chart"

DIRECTIONS:

1. Have the students put two ice cubes in each of the ten cups.
2. Put a pair of cups (see the "Freezing and Melting Chart") in each of five locations. Wait 15 minutes. If liquid has formed, take a temperature reading and record. Leave the cups alone for 20 minutes this time, then take temperatures and record.

3. Using a glass measuring cup, record the volume of liquid in each cup.
4. Observe and record which cup had the greatest volume of water—insulated or plastic? Which location had the greatest volume?

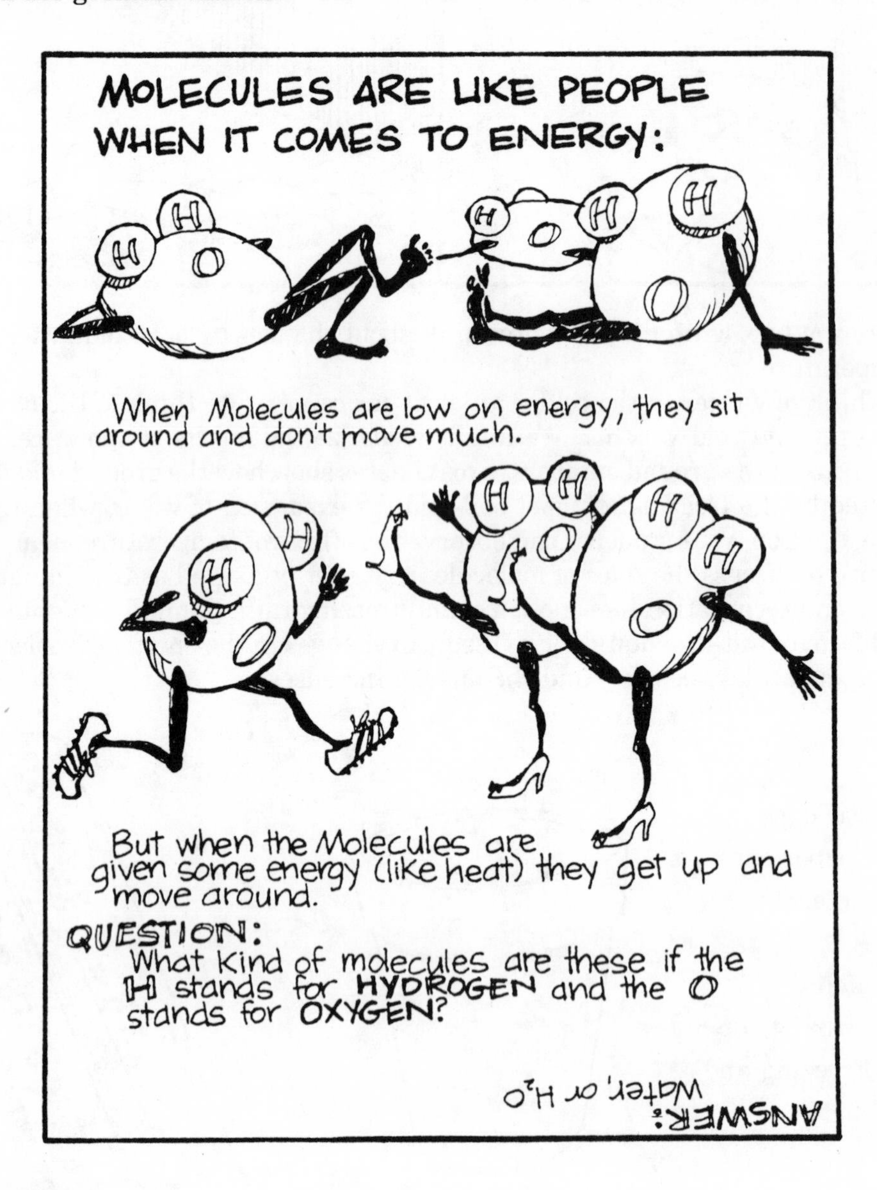

CONCLUSIONS:

1. What can your students conclude about the effect of location (light and temperature) on the melting of ice cubes?
2. What can your students conclude about how differences in insulation affect the melting of ice?
3. What kind of insulation is found in students' homes? Must water pipes be protected from freezing in your area? Discuss this with your students.

Name ______________ Date ______________

FREEZING AND MELTING CHART

LOCATION	TYPE OF CUP	DESCRIPTION AFTER 15 MINUTES	DESCRIPTION AFTER 20 MINUTES	TEMP.
Refrigerator	insulated			
	not insulated			
Outdoors	insulated			
	not insulated			
Window with Sun	insulated			
	not insulated			
Closet	insulated			
	not insulated			
Heater/ Radiator	insulated			
	not insulated			

W6: Observing Crystals

Winter is a time to observe and study crystals, especially as they appear in beautiful snowflakes. Even if you live in a warm area, this activity will help your students understand the formation and complexity of crystals, as well as the characteristic of uniformity all crystals share.

Crystals provide information about cloud types, temperatures, and heights of clouds. The shape of crystals varies—many are six-sided or hexagonal in shape, but other patterns exist. After doing this investigation, have your students read about crystals in the library. You may want them to do follow-up mini-reports on crystals and draw some pictures of crystals. Here they will be growing crystals from a solution of sugar and water. The sugar is the seed, or catalyst, that starts the crystal formation.

CAUTION:

Because boiling water is used in this activity, adult supervision is absolutely necessary!

EQUIPMENT:

Glass thermal-treated measuring cup
10-inch long string
Spoon
Pencil
Sugar
Boiling water (CAUTION: To be poured only by an adult)

DIRECTIONS:

1. Pour boiled water into a glass measuring cup. Add 1/2 cup of sugar and stir to dissolve.
2. Tie the string to the pencil and put it in the cup so that it dangles in the sugar water. Wait and leave undisturbed for four days.

CONCLUSIONS:

1. Have your students tell what happened to the string and describe what they saw.
2. Have the students tell what happened to the water.
3. Have your students read and find out what happens inside a cloud when crystals are formed. They will see that all crystals have a definite pattern or symmetry about them. The ions or atoms are arranged in a specific pattern, and this is a characteristic of all crystals.

W7: Examining Shadows

This activity can be done several times during the winter season, and throughout the year. It gives the students an opportunity to compare their findings and increases the reliability of their observations.

Don't miss the opportunity to do this activity on the first day of winter, the Winter Solstice. Discuss the fact that the sun's position is lower in the sky in winter. The shadows are longer than they would be if this activity were done in the summer.

Have the students measure their shadows at different times of the day—morning, noon, and late afternoon. Compare the lengths of the shadows at different times, noting the position and brightness of the sun and the role of the clouds.

Use the chart to record the length of the shadow and the time the measurement was taken. Discuss findings for different times of the day.

EQUIPMENT

2 stakes to mark the shadow

Long rope (jump rope) or string

Yardstick

Pencil

Shadow chart

Sunshine

DIRECTIONS:

1. Mark the beginning and ending positions created by the shadow. Measure and record results on the chart using the rope and yardstick.
2. Repeat this at three different times and discuss your findings.

W8: Observing GEOTROPISM

Often during the winter, many plants do not get enough sunlight and water, or the appropriate nutrients for growth. This is an excellent time to introduce your students to the effects of *geotropism*.

Geotropism is the process by which plant roots grow downward (positive geotropism) or in a different direction (negative geotropism). Roots should grow down toward the center of the earth, influenced by the pull of gravity. Regardless of how seeds are planted, stems grow upward and roots downward. There are certain chemicals inside plants that react to light and the pull of gravity.

This investigation can be done as a demonstration for the entire class. You will need two indoor plants of equal height, potted in identical flowerpots. The plants should be about five inches tall. You will also need a box large enough to cover both plants. The box should have no holes in it. (See the illustration below.) There is an excellent science video listed in the Appendix that correlates with this lesson.

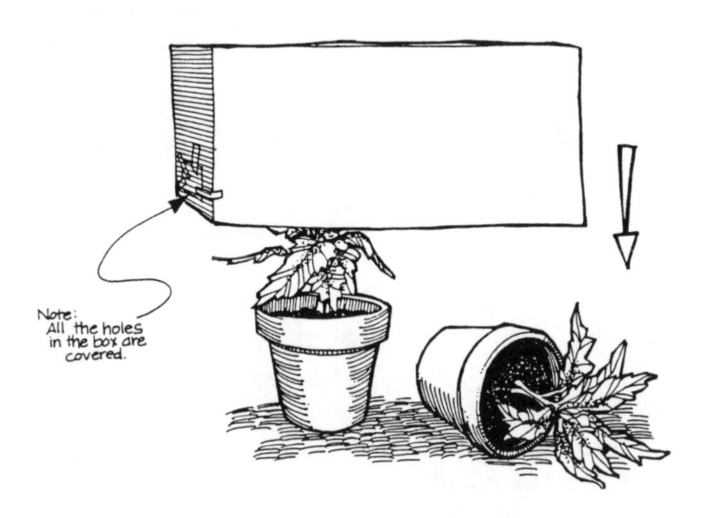

DIRECTIONS:

1. Moisten the soil of the plants.
2. Find a location where the plants can rest for four days or more.
3. Place them on the table (or shelf or floor). Put one upright in a normal position, the other on its side.
4. Put the box on top of the two plants. Leave undisturbed. After four days, remove the box. Record observations—you may want to make a poster for the classroom or a bulletin board of the results.
5. Carefully remove the plants from the pots and gently shake away soil to observe roots. Describe. Repot plants and give water.
6. When plants are deprived of light, what can happen to their leaves? The best kind of light for most green plants is sunlight. Without it, plants may die and their leaves may grow pale and thin.

W9: Understanding the Aurorae

The Great Aurora is best viewed on a clear, dark night like the ones we have in January in the Northern Hemisphere. This investigation involves gaining an understanding of the principles of magnetism and the earth's magnetic field. This can then be related to the northern and southern lights of the Aurora Borealis and the Aurora Australis.

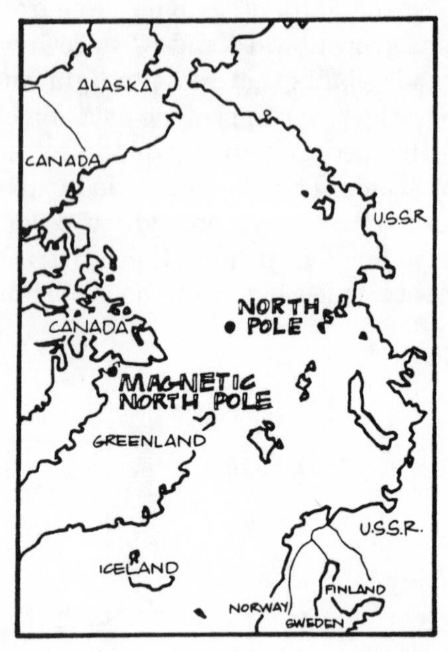

The Aurora Borealis is the light phenomenon viewed in the night sky best seen near the arctic region of the earth (the north magnetic pole, 78.6° North latitude, 70.1° West longitude).

The Aurora Australis is best viewed in the Southern Hemisphere near the south magnetic pole (78.5° South latitude, 106.8° West longitude).

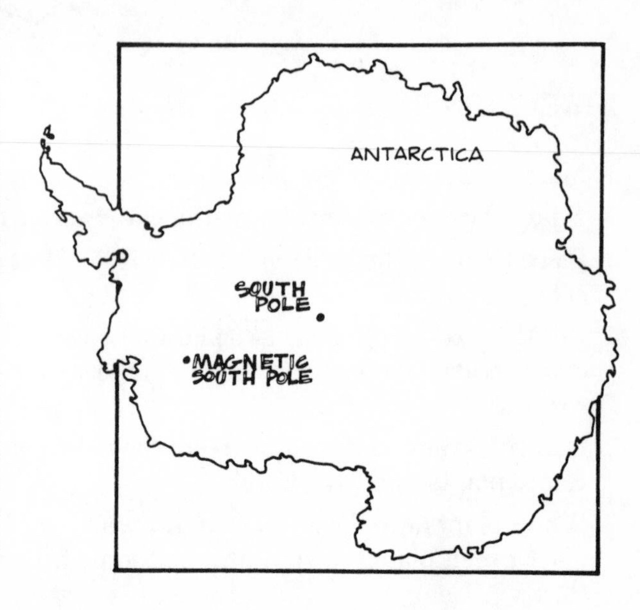

Aurora means "dawn" in Greek. Scientists study the sun and its storms to determine whether an aurora might reach the earth. The intense temperatures of the sun and its gases cause solar flares. The gases escape the sun's outer surface or chromosphere. The particles (mostly protons) meet with the earth's upper atmosphere of atoms and molecules of gas (like oxygen and nitrogen). The collision of the atmospheric particles creates surplus energy, as well as radiation and light in the aurora's colors.

In winter the earth is moving faster than it does in other seasons, and the charged particles are attracted to its magnetic poles. The following investigation will help your students understand magnetic attraction and the strength of magnetic poles. Relate findings to the Great Aurorae. (See the suggested video list in the Appendix.)

Name ______________________ Date ______________________

W9: Understanding the Aurorae

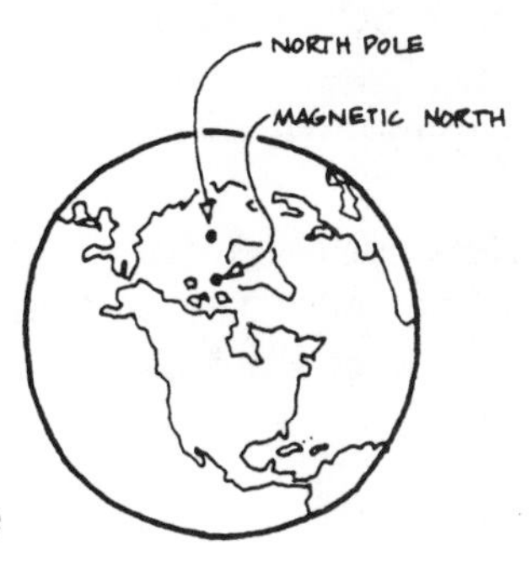

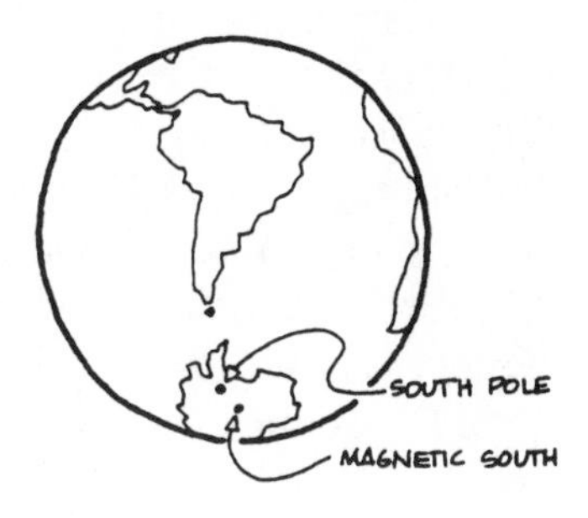

EQUIPMENT:

Iron filings
2 bar magnets
Paper
Plate of glass (CAUTION: To be used with adult supervision)

DIRECTIONS:

1. Sprinkle the iron filings on the paper.
2. Take one magnet and hold it under the paper and observe the movement of the iron filings.
3. Do the same with the iron filings on the plate of glass.
4. Repeat steps 1 through 3 using two magnets this time. First, touch the opposite ends (north and south poles). Observe the filings. Then touch the same ends (north and north; south and south). Do this on paper and on glass.

CONCLUSIONS:

1. Did you observe any patterns created around the ends or poles of the magnets? When did this occur?
2. How did the filings react to the magnets of the same poles? Of different poles?
3. Discuss the role of the earth's magnetic field and the attraction of the sun's charged particles. How does this create the luminous phenomena in the sky called the Great Aurorae?

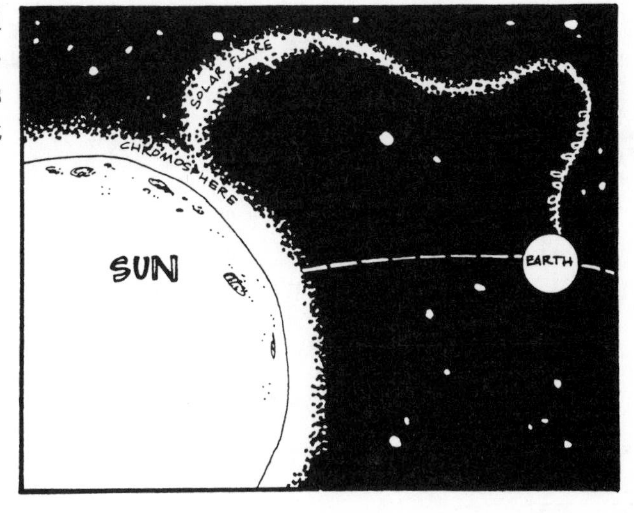

W10: Going on a Winter Scavenger Hunt

This hunt may be done in school or at home. It may be done on an individual or a small-group basis. Teams with five or six students and a captain may be formed.

Have the students create a few extra items to be added to the list prior to beginning the hunt. Discuss the fact that many of the items are to be located but not returned to the classroom. Identification counts!

Many of the items relate to investigations in this section, such as the ones on animal tracks and homes. You may want to follow up with the enrichment activities found in the WINTER SEASON ENRICHMENT SECTION; for example, the song "Tracking," Writing Fables on Tracking, and Making Wildlife Casts.

Reproduce the list and give your students copies prior to the hunt. Have fun!

WINTER SCAVENGER HUNT LIST:

1. ANIMAL TRACK *
2. EVIDENCE OF AN ANIMAL'S WINTER HOME *
3. DEAD LEAF (from a deciduous tree)
4. FROST CRYSTALS *
5. EVIDENCE OF FROST HEAVE *
6. EVERGREEN LEAF (needle)
7. EVIDENCE OF STATIC ELECTRICITY *
8. WOODPECKER *
9. LICHEN *
10. YOUR STATE BIRD *

* LOCATE, BUT DON'T PICK OR TAKE

Name ______________________ Date ______________________

W11: Examining Winter Weather

Meteorology is the branch of science that deals with the study of the earth and its atmosphere, especially in relation to weather and weather forecasting. Using the national/world weather report chart, read and examine the forecasts for the cities listed. Look at the different highs and lows in the columns and at the next-day forecast (February 10). Then answer the questions regarding information provided on the chart and the line graph. You may want to follow up at home by consulting your local newspapers for similar graphs and charts.

QUESTIONS:

1. Which city in the U.S. had the highest temperature on February 9?

____

2. Which city in the U.S. had the lowest temperature on February 9?

____

3. Which cities in the U.S. and world were expecting clear weather for the next day?

____

4. Which cities were expecting rainy weather? ______________________

____

5. Use this line graph to answer this question. Which month shows the highest amount of precipitation?

____

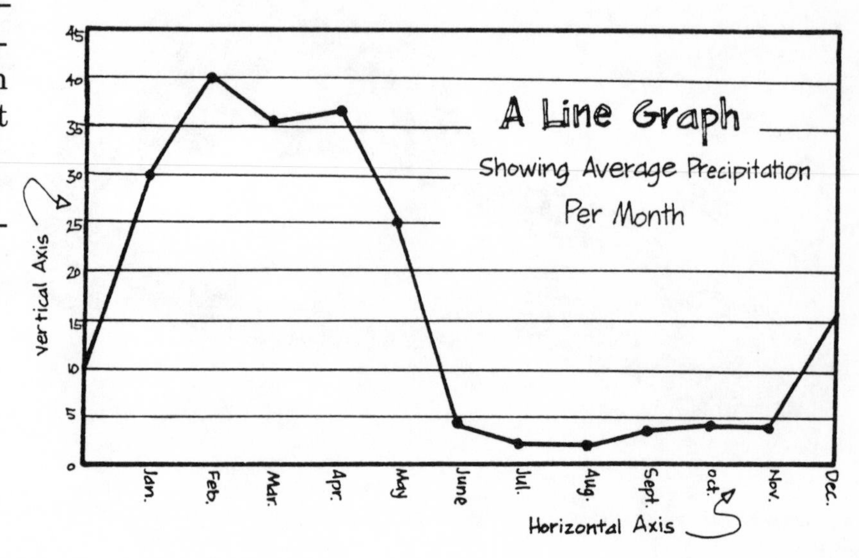

A National/World Report of Weather Forecast for February 10

Name of city	High Temp. for Feb. 9	Low Temp. for Feb. 9	Precipitation for Feb. 9	Outlook for Feb. 10
Atlanta	68	59	1.94	rainy
Austin	77	49		clear
Baltimore	48	36	.16	cloudy
Buffalo	49	34	.30	rainy
Charleston	75	62		cloudy
Chicago	55	37	.01	cloudy
Detroit	49	33	.29	cloudy
Honolulu	85	72	.02	cloudy
Indianapolis	62	50	.42	cloudy
Los Angeles	64	55	.35	rainy
Milwaukee	37	32	.47	cloudy
New Orleans	74	55	.10	cloudy
Oklahoma City	54	40		clear
Phoenix	66	49	.23	cloudy
Raleigh	75	48	.85	cloudy
St. Pete/Tampa	88	74		cloudy
Seattle	53	39		cloudy
Washington	48	42	.14	cloudy

World				outlook for Feb 10
Amsterdam	40	35		cloudy
Athens	57	49		cloudy
Bangkok	90	80		clear
Barbados	86	74		clear
Berlin	40	32		cloudy
Cairo	77	58		cloudy
Geneva	38	33		cloudy
Jerusalem	68	48		cloudy
London	45	35		rainy
Montreal	32	18		clear
Moscow	32	27		cloudy
Paris	35	28		cloudy
Vienna	46	33		cloudy

W12: The Winter Sky—Looking for Orion

Similar to activity F12 on Pegasus, have the students make constellation viewers with oatmeal boxes to view Orion. You may also make a glass peg mount slide to put in your slide projector so your students could see Orion on the screen, similar to projections in a planetarium. Two Orion patterns are given—one is for the slide mount, the other is for the constellation viewer like the one made in F12. Directions for the glass slide peg mounts are below. (See page 179 for ordering the peg mounts.)

Orion contains seven of the twenty brightest stars in the sky. It also contains two famous nebulas. One of its stars is called "Rigel," the seventh brightest star in the sky.

DIRECTIONS FOR PEG MOUNT SLIDES:

1. Take a role of 35mm film and expose it to sunlight outside. Unroll the film to do this.
2. Bring the exposed roll inside and cut it to fit inside the glass slide mounts. Then remove it from the mounts.
3. Using the appropriate pattern for Orion, make holes in the film to duplicate the shape, location, and size of the stars. (CAUTION: You should do this for the students, using a pin or compass.)
4. Put this perforated film back into the mount and insert into a slide projector for viewing. Darken the room, turn on the slide projector, and focus the constellation for viewing.

PATTERNS FOR ORION

Use this pattern if you are using a glass slide peg mount:

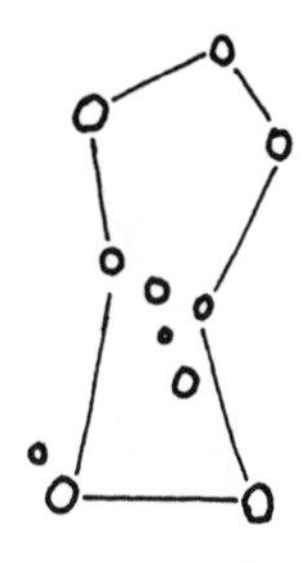

Use this pattern if you are using an oatmeal box:

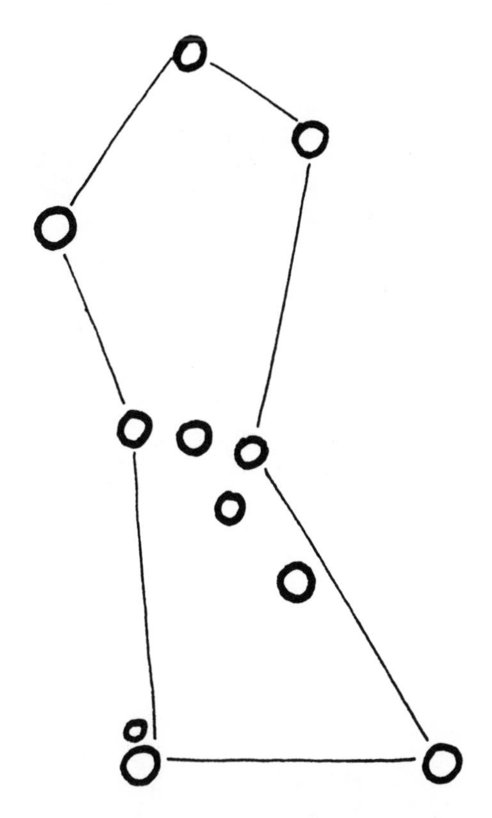

WINTER
enrichment
activities

Winter Cookery—Making a Mini-Solar Food Warmer

EQUIPMENT:

1 no. 10 cafeteria can
1 empty coffee can
Packing material (newspapers, styrofoam, plastic bubbles)
Plastic lid for coffee can
Scissors
Plastic food wrap
Dark construction paper
Tape
Sandwich bag
Doughnuts or other food for heating
Rubber band

(Special thanks to Jonathan Craig, Director of Ecology at Talcott Mountain Science Center in Avon, CT for this activity.)

DIRECTIONS:

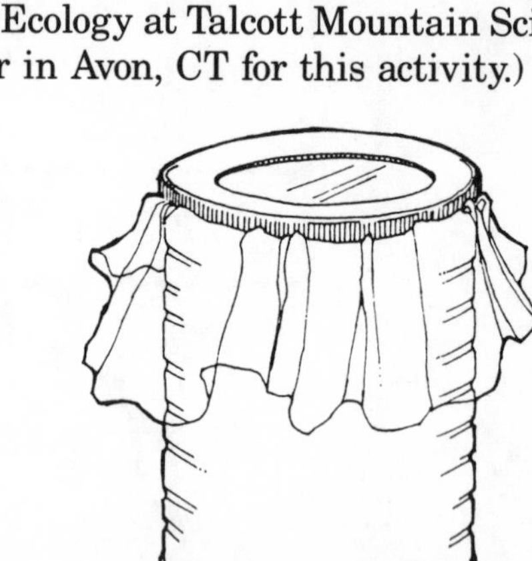

1. Line the inside of the empty coffee can with the dark paper, and tape paper to hold it in place.
2. Cut out the middle circle of the plastic lid and cover the top with plastic food wrap. The lid will hold the food wrap in place. Use tape if needed to secure. (See the illustration.)
3. Put the doughnut in a sandwich bag and place it inside the coffee can.
4. Put the lid on the coffee can and place the can inside the cafeteria can. Fill the space between the two cans with your packing material. (You and your students may want to try different types of insulation.)
5. Put another piece of the plastic food wrap on top of the two cans and hold it in place with a rubber band.
6. Put the mini-solar food warmer in the sun and angle it to catch direct light.

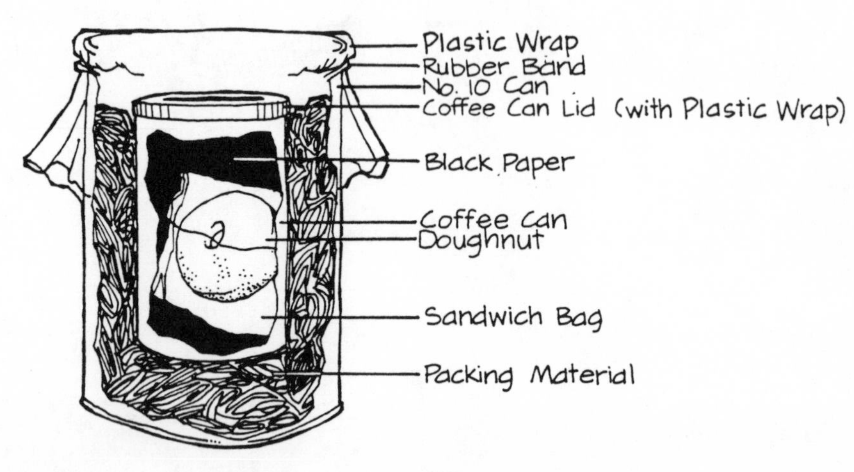

Making a Pocket Sundial

A sundial is a device used to show the time of day based upon the shadow caused by the dial's pointer, or gnomon. On certain days of the year, your sundial will be most effective—on the Winter and Summer Solstices and on the Vernal and Fall Equinoxes. Try this activity four times a year or more, and try it at noon.

(Special thanks to Jonathan Craig, Director of Ecology at Talcott Mountain Science Center in Avon, CT for the pattern.)

EQUIPMENT:

- Sundial pattern
- Colored cardboard or card-stock paper
- Markers
- Thread
- Tape
- Glue
- Watch
- Mechanical compass to punch holes (CAUTION: To be used only under adult supervision)
- Protractor (optional)
- Scissors

DIRECTIONS:

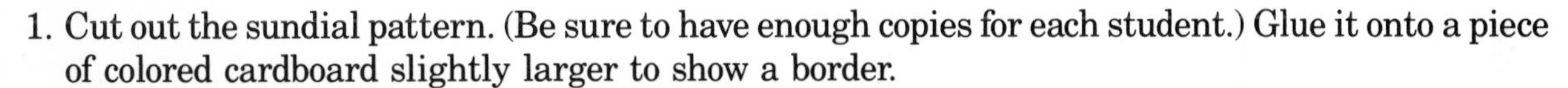

1. Cut out the sundial pattern. (Be sure to have enough copies for each student.) Glue it onto a piece of colored cardboard slightly larger to show a border.
2. Bend the glued sundial along the dotted line at a right angle, as shown here.

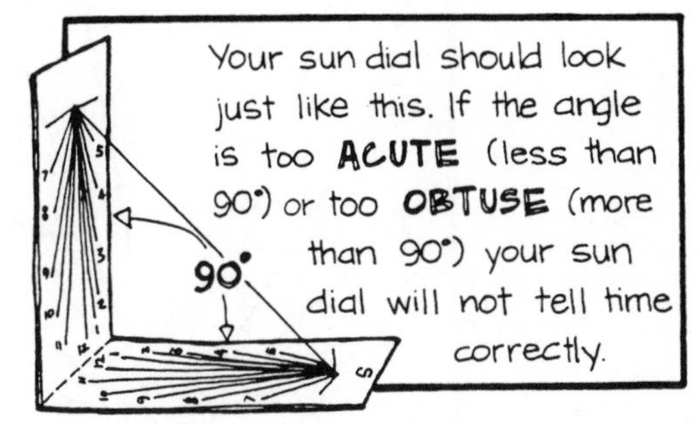

3. WITH TEACHER SUPERVISION, take a mechanical compass and point it downward, away from your eyes. Punch a small hole in the sundial as indicated on the pattern. Punch it where the snowflake-like symbol is shown, at both the top and bottom of the pattern.
4. Thread a four-inch piece of thread through the holes and tape it taut so the right angle is maintained.
5. Wearing a watch to regulate your dial, go outdoors on a sunny day. Expose the dial to the sun so the gnomon (the thread) faces the North Star. Open your card so the thread casts a shadow on the card. It will appear as a fine line shadow on one of the numerals representing the time of day.
6. To set the gnomon, look at your watch. At 12 noon, the shadow of the gnomon should be on the numeral 12; at 2 o'clock, the shadow of the gnomon should be on the numeral 2 on the dial. Once set, place the card in its right angle position on the ground and put a rock on the bottom of it to keep it from blowing away. Observe it as the shadow moves. Check it after an hour or more. Every degree it moves represents 4 minutes of time.

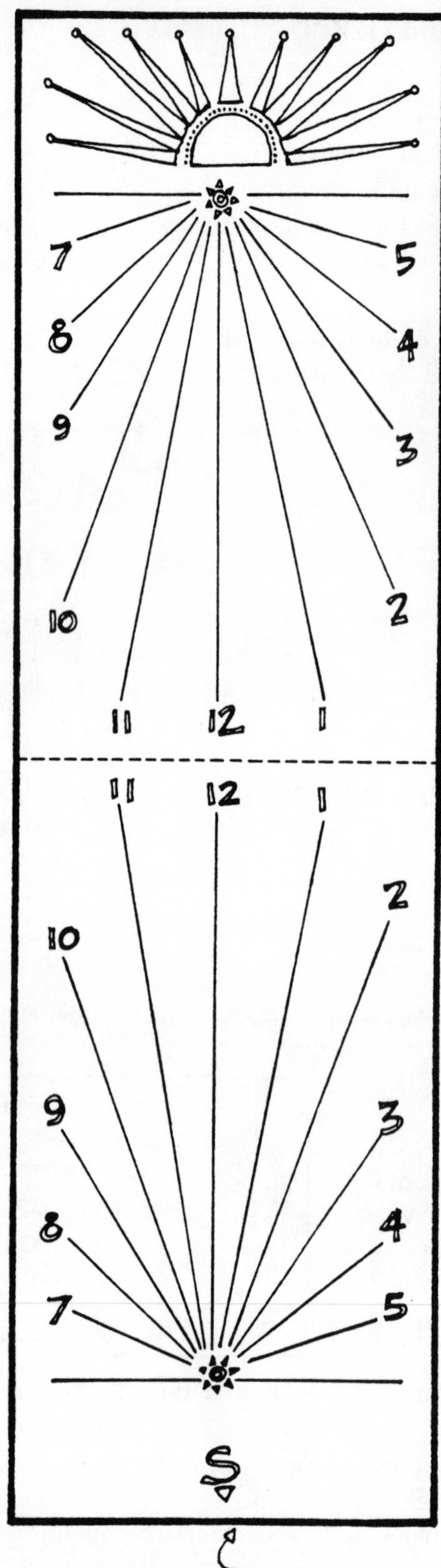
7
5
8
4
9
3
10
2
11
12
1
11
12
1
2
10
9
3
8
4
7
5
S
BEND AT
DOTTED LINE
CUT HERE
= THREAD HOLES

Making Casts of Wildlife Prints

MATERIALS:

Wildlife footprints to cast (find these outdoors)
Spray mister with water
Aluminum foil
Scissors
Cardboard
Spade
Acrylic paint
Paintbrush
Bucket
Petroleum jelly
Masking tape
Plaster of paris (read mixing instructions on bag)

DIRECTIONS:

NOTE: Since plaster of paris sets quickly, prepare other materials ahead of time.

1. Take the cardboard and make a strip collar or frame with strips to fit around the print. Smear petroleum jelly on the inside of collar or strips. (The petroleum jelly assists in removal later.) Cover and tape aluminum foil around the cardboard as shown.
2. Mix the plaster of paris with water to a thick milkshake consistency. Do this outdoors as the mixture sets quickly.

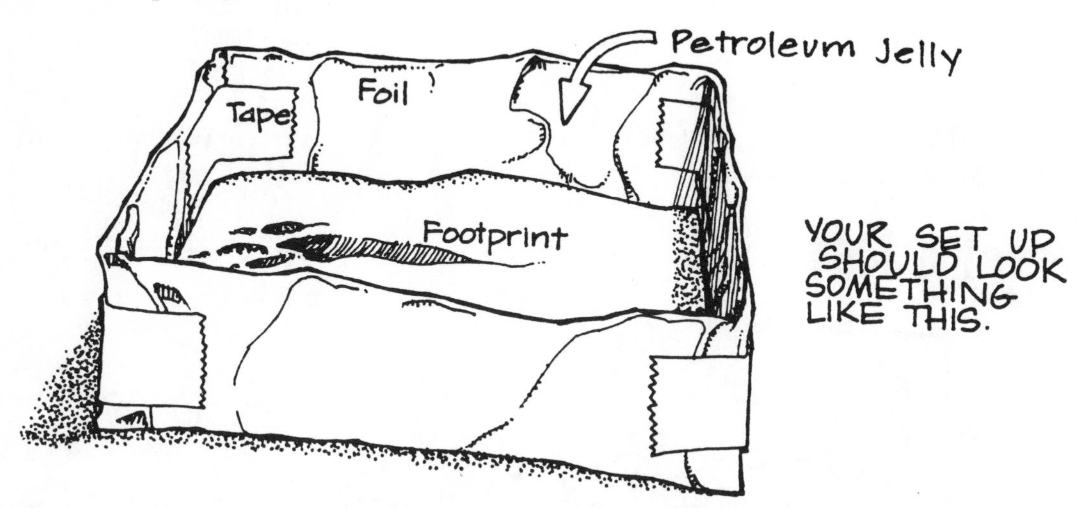

3. Mist the footprint with water. Put the cardboard around the print to box it in. Pour the plaster of paris inside the frame. Do not overfill. Put rags around the outside of the cardboard collar to keep plaster from running outside collar.
4. Allow time to dry. Return indoors if necessary. Drying time varies, depending on weather and temperature. Use the spade to lift the cast. Clean it back in the classroom and paint it with acrylic paint. If desired, reverse this impression by using your cast to make a footprint inside a fresh frame of plaster of paris outdoors. Remove the cast and let it dry.

Name________________________ Date____________________

Winter Science Stories, Songs, & Poetry

— ANIMAL TRACKS : WHO DONE IT? —

Write a fable about the animal tracks found in the winter season. Explain whose tracks the footprints are and how they got there.

Make the plot of the story tell a moral or truth to the reader or listener. (For example, the Fable of the Tortoise and the Hare tells us "Slow and Steady wins the race.")

OR

Write a Mystery Story about the animal tracks.

Name ____________________ Date ________________

WINTER SCIENCE STORIES, SONGS, & POETRY

1. I see footprints in the mud, frozen like a plaster cast.

Wonder whose footsteps these could be, impressions made to last.

2. I see footprints in the snow,
deep in the fluffy white flakes.
Some are large, yet some are small;
a detective I do make.

3. I see footprints in my house,
upon my mother's polished floor.
Mud and snow stuck to my boots;
I scurry out the door.
I scurry out the door!

SUGGESTIONS:

Percussion Accompaniment – Orff Instruments

Music by:
Patricia McCamish Donohoe

Lyrics by:
Julia Moutran

Name ______________________ Date ____________________

Winter Science Stories, Songs, & Poetry

WRITING ACTIVITIES: USING THE NEWSPAPER

Cut out several newspaper headlines about the weather. Write a paragraph to summarize your findings.

Create a travel advertisement for a winter vacation in an area of the United States.

BE A REPORTER!

Write an eyewitness report about the weather in your area. Answer the "five W's"—Who, What, When, Where, & Why. Include an interview, quotations and a headline.

Name ____________________ Date ____________________

WINTER CROSSWORD PUZZLE

ACROSS

2. Greek word for "Dawn"
5. The growth of roots with respect to the force of gravity
7. The type of root system made up of many roots

DOWN

1. German word meaning "Time of Water"
3. Name of the seventh brightest star in the sky
4. Name for the first day of winter; the shortest day of the year
6. Type of root system made up of one main root and many smaller, hairlike roots
8. One of the largest constellations in the winter sky—named for the "whale"

Snowflakes

Using the poem "Snowflakes," have the students read about snowflakes and crystals. Even if you live in an area that does not have snow, children everywhere are fascinated by snow and snowflakes and the formation of snow.

Have them read about clouds and cloud types, weather forecasting, blizzards, and snowflakes. Winter is an excellent time to introduce students to the study of crystals. Crystals provide information about the types of clouds, temperatures, and heights of the clouds. There are also some excellent science videos on clouds and weather to reinforce this.

You might also want your students to write their own poems about the snow, weather, crystals, or clouds after reading and sharing "Snowflakes."

Name ______________________ Date __________________

WINTER SCIENCE STORIES, SONGS, & POETRY

SNOWFLAKES

Snowflakes falling to the ground,
Whiten places dark and grey.
Frozen crystals from the sky
Drifting by my way.

Catching you is lots of fun –
So is counting all your rays:
Atoms fixed in frozen form,
Carved in countless ways.

Little flakes stick to the ground,
And for a while there you'll stay
'Til you'll be used to make a man
Of snow this very day.

Name ______________________ Date ______________________

Feeding the Birds

In the winter, birds often need extra food for fuel and warmth. Suet is an important food source for them. Here are three ideas for you to make to feed the birds in your area.

1. Make a coconut or grapefruit bird feeder. Fill it with birdseed and suet and hang it on a tree.
2. Make a pine cone bird feeder. Cover the pine cone with peanut butter and bacon grease and roll the pine cone in birdseed. Hang the pine cone by a string or colored yarn and watch the birds gather to eat!

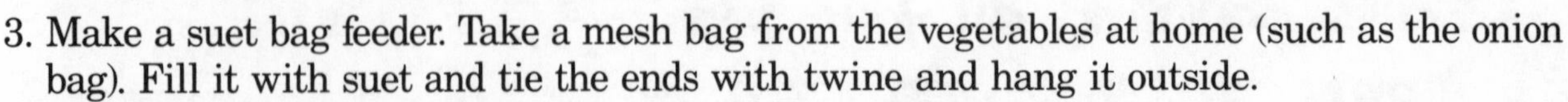

3. Make a suet bag feeder. Take a mesh bag from the vegetables at home (such as the onion bag). Fill it with suet and tie the ends with twine and hang it outside.
4. Be sure to hang the feeders high on the tree so other animals do not get the birds' food.
5. Make a "Bird Notebook" by completing the two charts.

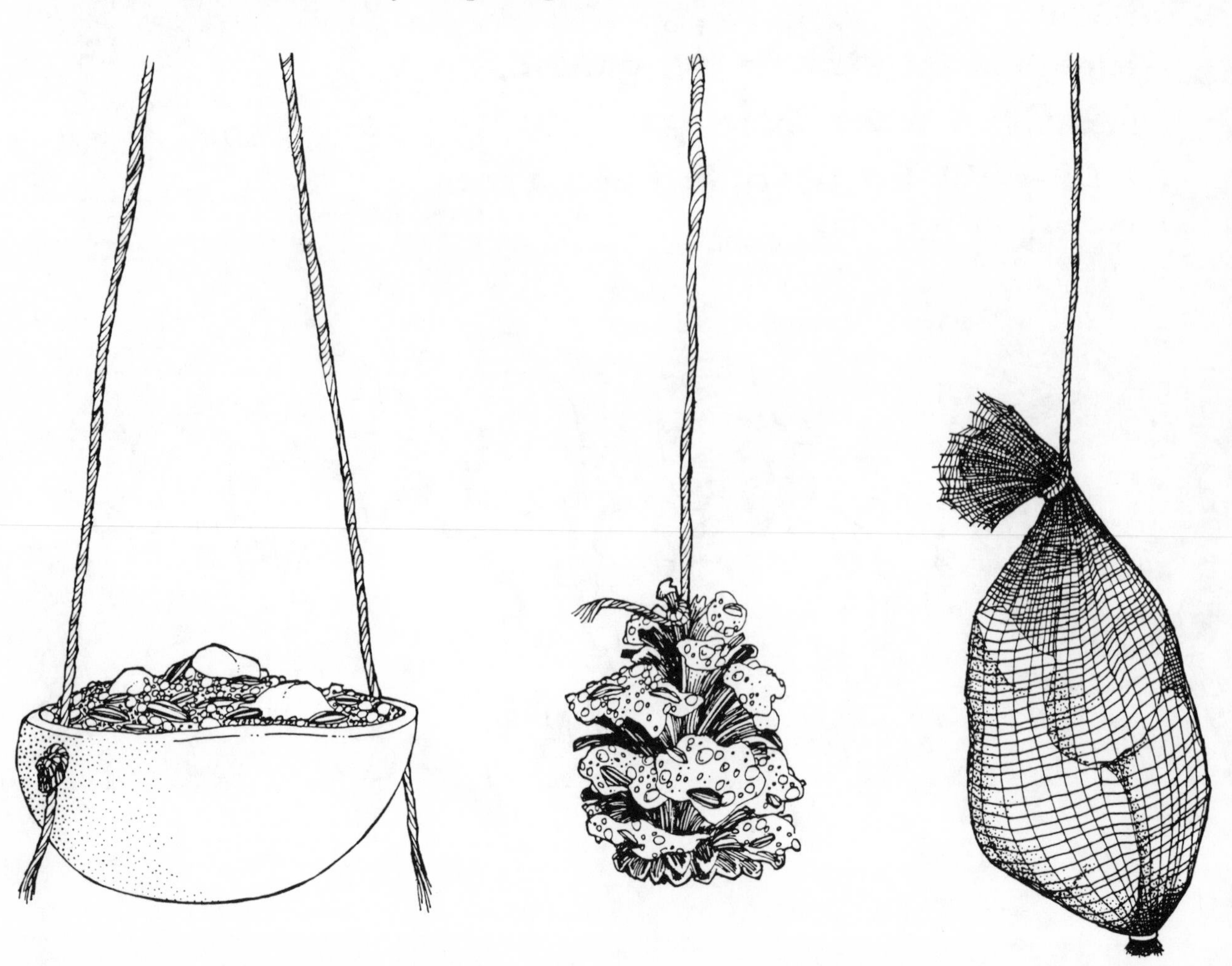

MY BIRD NOTEBOOK

This is to certify that I have fed and observed the following birds near my home or in my schoolyard.

The foods the birds have eaten include the following: (Check off any of these items)

_ Bird Seed
_ Peanut Butter
_ Suet
_ Fruit
_ Raisins
_ Thistle
_ Egg Shells
_ Crackers
_ Cooked Potatoes

_ Peanuts
_ Pumpkin Seeds
_ Sunflower Seeds
_ Meat
_ Bacon Grease
_ Cracked Corn
_ Toast, Bread, Muffins
_ Cereal
_ Berries

List any other foods here

____
____
____
____
____
____

This certificate has been completed by __________
Date __________

My Observations of Birds and Their Eating Habits

Date	Type of Feeder	Food in the Feeder	Number of birds at the Feeder	Type of Birds

The Spring Season

TEACHER'S GUIDE FOR SPRING

This is the third section of the book, representing activities for the spring season. Similar to the first two sections, there are twelve major activities (SP1 through SP12). These are followed by nine enrichment activities, including more science crossword puzzles, more science cookery, and a science song for spring.

Continue to use flow charts to monitor progress and send home the parent letter. Some excellent science videos that correlate with the activities are recommended in the Appendix.

Students can continue their exploration of the constellations in the sky. In the spring, they will focus on Leo with their constellation viewers, as well as learn about other suggested constellations in the spring sky.

The activities in Section Three allow students to learn about the composition of plants and the parts of flowers. They will have the opportunity to propagate plants by two different methods (see SP2 and SP3). They will continue to have fun on the Spring Scavenger Hunt and by examining spring trees. Comparisons with their findings in the other two seasons will help students understand the changes in their environment season by season.

Begin Section Three with the full-page illustration on page 84. Reproduce it for a science notebook and discuss these concepts with your students:

- Bulbs are blooming.
- Trees and leaves have changed since winter.
- Children's clothing indicates temperature change.
- Cirrus clouds appear in the sky.
- New leaf growth and blossoms appear on trees.
- Children are preparing a garden.
- Tadpoles are in the brook.
- Animals (sheep and those underground) now have babies.
- New insects and larvae appear in the spring.

You may also want to order *Collecting Bugs and Things* (see the Appendix) as a science activity storybook and unit for your classroom. It is filled with facts and activities about ladybugs, fireflies, butterflies, and moths. It includes a certificate for teaching children the care and feeding habits of insects in the spring and summer.

This spring section also has science spelling, writing science diamantes, and solar energy cooking. There is an activity on cool and warm weather crops, as well as suggestions for planting seeds with your students.

Just before your spring vacation, you may want to do the activities on spring weather, daylight saving time, spring tides, and spring temperatures (SP7 through SP10). Talk about choices in vacations and how the weather plays an important role in people's choices.

The activity on the Vernal Equinox and the first activity on "What and When Is Spring?" will help students understand the causes of the seasons—the position of the earth in its year-long journey around the sun; the position of the North and South Poles in relation to the earth's spinning on its axis. Compare the Vernal Equinox with the Autumnal Equinox. On both days, the length of day equals the length of night. Spring officially begins on the Vernal Equinox and fall begins on the Autumnal Equinox. The sun shines on the equator. These activities may be done as your first activities since they are fundamental to ushering in the first of spring.

Have your students make science notebooks on spring trees and flowers, including the history of their state flower. Enact a pressed flower exchange program with science pen pals in different states, similar to the Leaf Exchange Program in the fall.

Spring is a time to share the new life of flowers, trees, insects, birds, and other animals, and to appreciate the wonders of nature. These activities and your direction will help your students be aware and appreciative of these developments in their world.

Dear Parents,

Spring has arrived! In our science program, we will be studying about the world of nature during spring. During this time, your child will be investigating the following topics:

Changes in spring trees, leaves, and flowers;

Changes in the weather (spring temperatures);

The arrival of the Vernal Equinox and Daylight Saving Time;

Continued observation of the sky, including the constellations Leo, Virgo, Boötes, and Centaurus;

Observing new forms of life (insects, plants).

As your child learns more about his or her world of science, it is hoped you will share in these discussions and findings. Spring is a wonderful time of the year to participate together in the beauty of nature.

Thank you for your support of our science program.

Sincerely,

____

Teacher

SP1: What and When Is Spring?

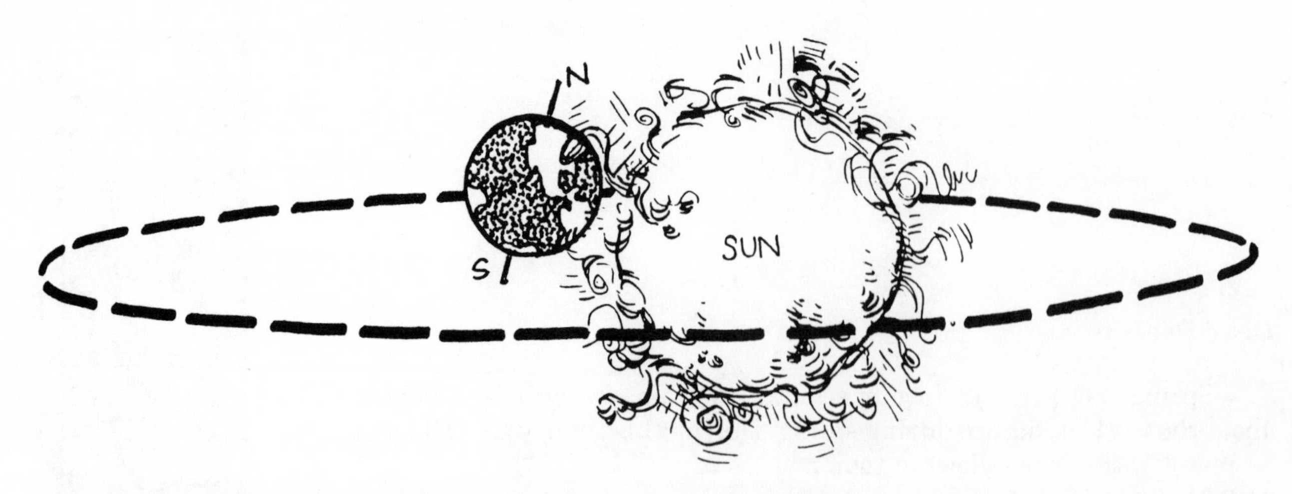

Each season should be introduced with a review of the position of the earth and the sun during the four seasons. Refer to activities F1 and W1 for review, especially the illustration for F1, "Here's What You're Acting Out."

Spring officially begins on the day of the Vernal Equinox, on or about March 21. The earth is orbiting the sun in its revolution. When spring comes, the sun moves higher in the sky than it did in the winter. Days begin to lengthen and the sun shines more directly on the earth, culminating with the longest day of the year on June 21.

Have your students make charts or posters depicting these concepts. This will help them understand how the earth revolves around the sun and why the seasons occur. Doing the "What and When Is Spring?" activity will also help them examine the position of the earth and its revolution around the sun during the spring season. They will see the tilting of the earth and the position of the poles, as well as its actual path during the next three months.

Name ______________________________ Date ________________________

What and When Is Spring?

EQUIPMENT:

Illustrations of the sun and earth (see below)
Glue
String or yarn
Colored markers, crayons, or pencils
Scissors
Posterboard or construction paper

DIRECTIONS:

1. Cut out the illustrations of the earth and the sun at the bottom of this sheet.
2. Glue the sun in the middle of the posterboard or construction paper. Color it.
3. Take the model of the earth and think about the position of the earth in relation to the sun during the spring season. Where is it in its year-long journey around the sun?
4. Take the yarn or string and form an ellipse around the sun, with the sun in the middle of the ellipse. Glue the string to the paper. Be sure you make an *ellipse*, not a circle, to represent the path of the earth for one year.
5. Now, place your model of the earth in its proper position. Think about the placement of the earth on the elliptical path and the position of the North and South Poles. Remember, the earth is tilted on its axis. The tilting of the earth causes the changes in temperature and weather during the different seasons.
6. Color the earth.

CONCLUSION:

How does the tilting of the earth cause different seasons?

____

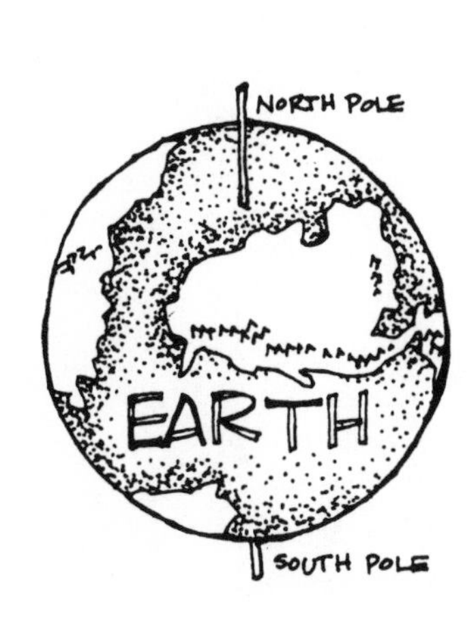

Name ____________________ Date ____________________

SP2: Growth Cycle of Plants

Write Four Things You Think Plants Need

1. ____________________

2. ____________________

3. ____________________

4. ____________________

In this activity, you will be dividing, and thereby increasing, the number of plants from a parent plant. This activity will show you one form of "propagation." Propagation means to divide the parent single plant into more pieces, thereby increasing the number of plants.

In the spring, many people plant seeds, start gardens, propagate plants, and transplant plants.

EQUIPMENT:

- 1 indoor plant (such as a coleus or philodendron)
- 1 or 2 clay pots
- Clear plastic bag large enough to fit over plant cutting
- 1 cutting knife (CAUTION: To be used only by adult)
- Bag of potting medium (or 1/2 peat moss and 1/2 perlite)
- Water and plant mister

Name ______________________________ Date ____________________

SP2: Growth Cycle of Plants (continued)

DIRECTIONS:

1. Soak the clay pot(s) in water for a few minutes.
2. Fill the pot(s) 3/4 full with potting medium.
3. Find the leaf node on the plant. The node is where the leaf and stem connect, as shown in the illustration.
4. IMPORTANT: The node should only be cut by your teacher or an adult supervisor. The plant should be cut beneath the node so that you can have a new cutting from the parent plant. Be sure the cutting has leaves on it.
5. Carefully place the cutting into the potting medium. Plant it carefully and moisten it and the medium with water.
6. Gently place the plastic bag on top of the plant to create a mini-greenhouse.
7. Place the plant in the sun, away from drafts.
8. Observe its growth over the next three weeks.
9. Mist and lightly water when needed.
10. What do you think will happen to the cutting? (This is your hypothesis.)
11. Did you guess any of these four things that all plants require? (Water, light, air, and proper temperature)

CONCLUSIONS:

1. Turn this sheet over and at the top, draw a picture of your cutting after three weeks. What has happened?
2. At the bottom of the other side of this sheet, draw a picture of the parent plant. What has happened?
3. Remove the cutting and gently remove the plant from its potting medium. Examine the roots. What has happened? (Was your hypothesis correct?)

Name ______________________ Date ______________________

SP3: Plants Grown from Roots

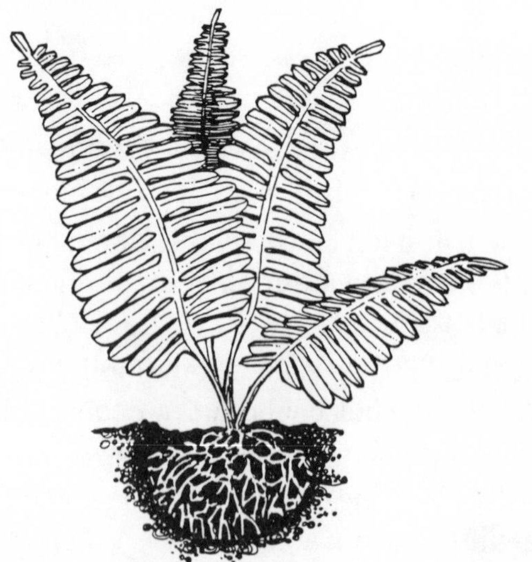

In activity SP2, you grew plants divided from a single parent plant by "propagation." In this activity, you will learn more about the root system of plants and propagating plants by root division.

Why does a plant have roots? Did you know they help bring water and minerals from the soil to the plant? See if you can find the difference between a "taproot" and a "diffuse root." What does the word "diffuse" mean? Write the definitions here.

TAPROOT: ____

____

____

____

DIFFUSE ROOT: ____

____

____

____

Name ______________________ Date ______________________

SP3: Plants Grown from Roots (continued)

EQUIPMENT:

1 fern plant
Potting soil
Water
Knife (CAUTION: To be used only by adult)
3 potting containers for replanting

DIRECTIONS:

1. Take a fern plant and carefully remove it from its container.
2. Gently shake the dirt to expose the root ball. Look at the hair-like projections called "root hairs."
3. Your teacher (or other adult) will take the knife and carefully cut the root ball, dividing it into three parts.
4. Replant all three parts of the plant in the containers with potting soil.
5. Water the plants and place them in the same location. Give them adequate light and keep them away from drafts.
6. Make a guess (a hypothesis) about what will happen to the plants both above and below the soil level.
7. Observe and record the information of their growth.

CONCLUSIONS:

1. Describe the progress of the three plants one week after root division. ______________________

____

____

____

2. Describe the progress of the three plants two weeks after root division. ______________________

____

____

____

Name ______________________ Date ______________________

SP3: Plants Grown from Roots (continued)

3. Describe the progress of the three plants after three weeks.

____

____

____

____

4. After four weeks, remove the ferns from their containers and carefully remove the soil to examine the root balls. Describe your observations.

____

____

____

____

5. What happened? Was your hypothesis proven correct?

____

6. Did your fern have a diffuse root system (many roots) or one thick and large main root system (taproot)?

____

Name ______________________ Date ______________________

SP4: Examining Plants and Flowers

Spring is a great season to observe plants and flowers. So many beautiful plants are now flowering. You have learned about leaves, roots, and leaf nodes. Now you will learn about the parts of a flower. Not all plants have flowering parts. Sometimes a plant has a flower or bloom for only a short time.

The flower contains parts that produce the seeds of the plant. Look at the illustration below.

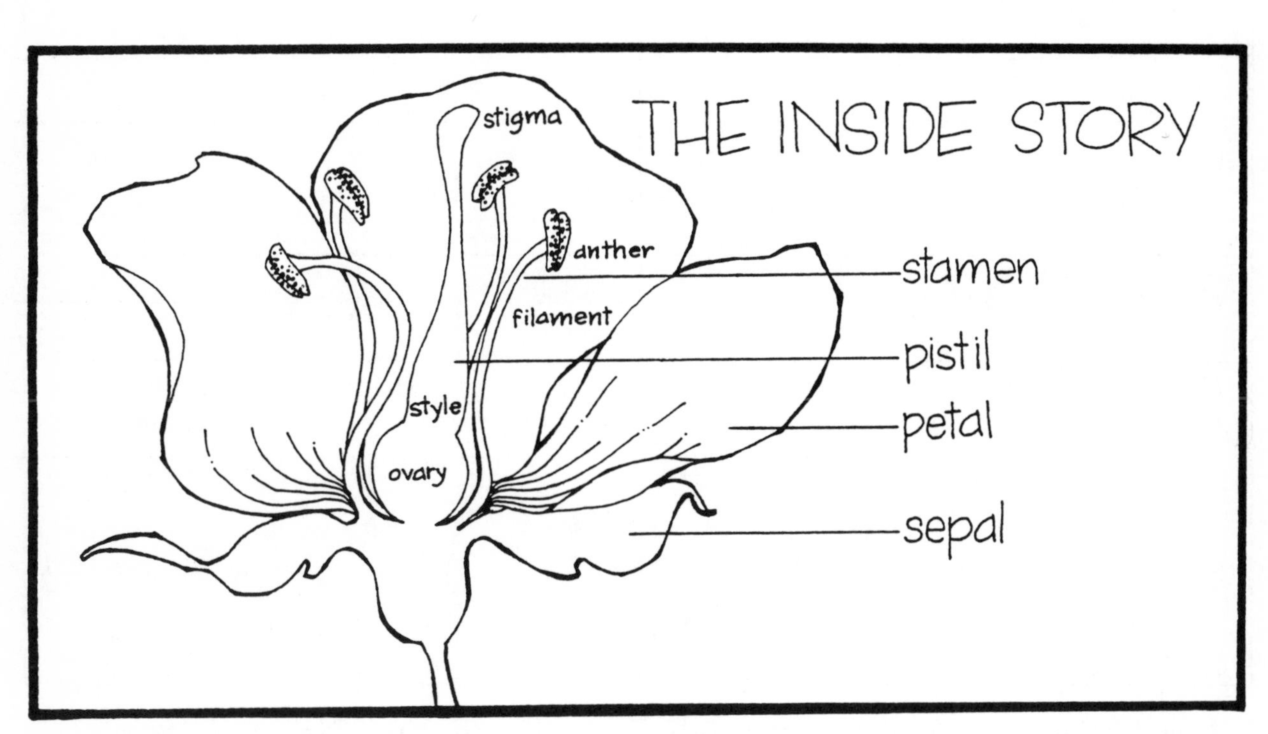

For this activity, you will need a plant with a flower to observe. A tulip, jonquil, daffodil, or any other type of flower available in your area that has easy-to-identify parts will do just fine.

EQUIPMENT:

1 spring flower
Scissors
Construction paper
Hand lens (CAUTION: Handle with care)
Pencil and crayons
Tape

Name ________________________________ Date ________________________

SP4: Examining Plants and Flowers (continued)

DIRECTIONS:

1. Look at the flower with your lens. Look at the petals. Record here the number of petals.

   ____

2. Look for the pistil, stigma, style, and ovary. Draw pictures of each one here.

PISTIL	STIGMA
(The center part of the plant contains the stigma, style, and ovary.)	(The tip of the pistil.)
STYLE	OVARY
(The slender part that connects the stigma with the ovary; it resembles a tube.)	(The base of the pistil, it contains the ovules that help form the seeds of the plants.)

3. Look for the sepals, the green parts of the flower near the base of the petals. Sepals help protect the flower before it opens.
4. Take your scissors and cut off the stigma. Tape it in the box next to your illustration.
5. Cut off the filament and tape it in the box. The filament is the stem or stalk part of the anther (the knob of the stamen that makes the pollen that the plant needs for reproduction).
6. Use your scissors to carefully cut open the ovary. Do you see the ovules, which are the seeds found in the side of the ovary? Put some in the box next to your illustration.

CONCLUSIONS:

Explain how insects (bees and butterflies) help pollinate flowers and plants.

Name ______________________________ Date ______________________

SP5: Examining the Spring Trees and Leaves

Spring is a wonderful time to observe the seasonal changes in trees. Many trees begin to produce beautiful flowers. Buds appear and various insects are busy carrying pollen from flower to flower.

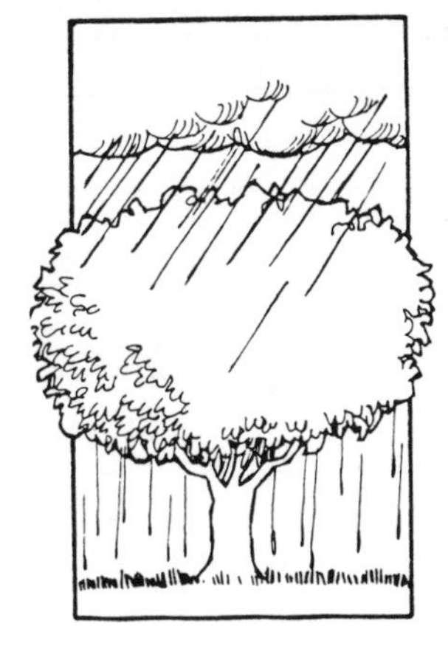

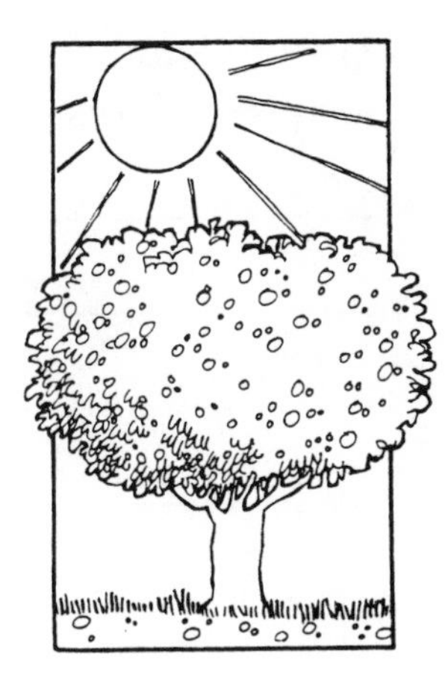

EQUIPMENT:

1 field guide on trees	Crayons
Pencil	Unlined paper

DIRECTIONS:

1. Go outdoors and look around you. Look for the seasonal changes in the trees.
2. Pick four different trees to examine.

3. Sketch the shapes of the trees on the back of this sheet.
4. Estimate the height of each tree. Write that information beneath each tree you sketched on the back of this sheet.
5. Look for evidence of buds or flowers. Describe any you find here. ______________________

____

____

6. Look for leaves and new growth. Describe the color of the new leaves. Touch and feel them, if possible. ____

____

____

____

Name ______________________ Date ______________

SP5: Examining the Spring Trees and Leaves (continued)

7. Describe any evidence of life around the tree. Look for homes in the tree or around the tree, footprints, insects, etc. Describe.______________________

8. Look at the roots and base of the tree. How old do you think the tree is? Guess.

____

9. Record the date of your observation. Use your field guide to identify the type of tree(s). Write the name beneath each tree you sketched on the back of this sheet.

____

	SUMMER DESCRIPTION	FALL DESCRIPTION	WINTER DESCRIPTION
Tree 1: ______________ (name)			
Tree 2: ______________ (name)			
Tree 3: ______________ (name)			
Tree 4: ______________ (name)			

10. On a sheet of unlined paper, draw a picture of one of the above trees during the spring.

CONCLUSION:

How do you think each tree will change during the four seasons? Describe how you think it will appear in the summer, fall, and winter. Use the chart above.

SP6: Going on a Spring Scavenger Hunt

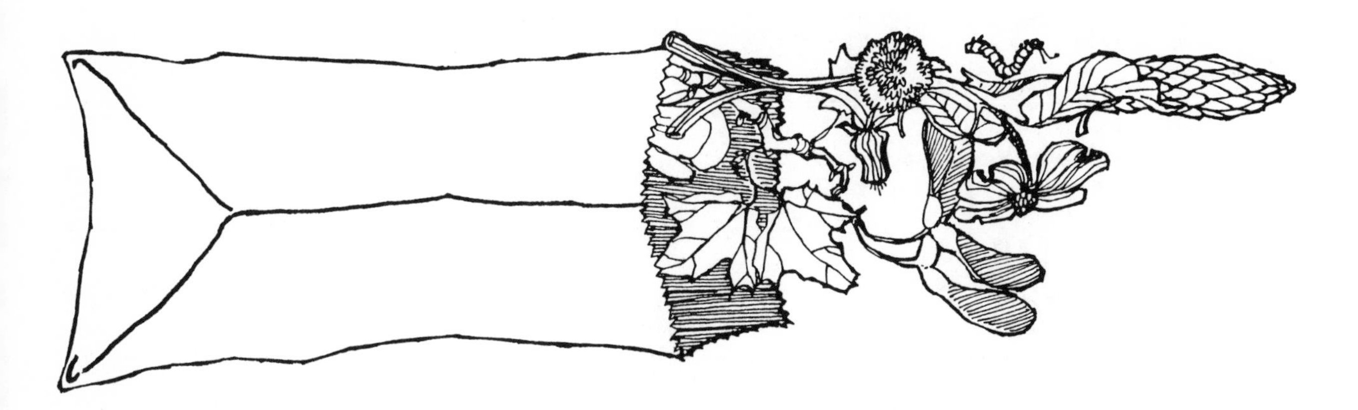

This Hunt may be done in school or at home. It may be done on an individual or a small group basis. Teams with five or six students and a captain can be formed.

Have the students create a few extra items to be added to the list prior to beginning the hunt. Discuss the fact that many of the items are to be spotted but not picked or touched. Identification and spotting count!

Students may want to do follow-up reports on their state flower and the flowers of other states. Each student could pick a different state and report on its state flower. Often the flower is in bloom at this time of the year. If leaf exchanges are done with other schools, this could be a good time to renew that project and prepare scrapbooks or collect seed specimens to exchange with another school. It's also a good time to talk about insects and pollen following your hunt.

Reproduce the list for your students and have fun! Don't forget that many of the correlated Coronet/MTI videos are excellent teaching resources for this activity and others in this book. Be sure to check the Appendix for references.

SPRING SCAVENGER HUNT LIST:

1. BUD ON TREE *
2. BIRD'S NEST *
3. COCOON *
4. ANT *
5. ANTHILL *
6. INSECT (other than ant) *
7. STEM WITH BUD SCAR *
8. YOUR STATE FLOWER *
9. DANDELION
10. FLOWER GROWN FROM A BULB *
11. SEED *
12. STAMEN *
13. PISTIL *
14. DEW *
15. POLLEN *
16. SPORE BEARING PLANT (mushroom) *

*LOCATE, BUT DON'T PICK OR TAKE

SP7: Studying and Measuring Spring Weather

This is a good time to have your students study the weather for four different cities during any spring day. Reproduce the map and the weather chart for each student. Have each student select four different cities in four different parts of the country.

They will collect weather information on the four cities using information from television, cable weather station channels, newspapers, and other sources (your local weather bureau). Talk about ways to find out about the weather conditions and why we should know about them. How does the drought in one area affect another area, for instance? You may want students to work in teams or small groups of four.

Plan a time to discuss their findings and have them share the findings with the class. Discuss differences in weather during the spring in different parts of the country. Do students have plans to vacation in certain areas during the spring vacation? Do you? Why do people favor certain vacation spots at this time of the year?

It's important to relate the findings about weather to the choices people make about their vacations, jobs, and places to live. This will help students understand the role of weather in their everyday lives.

SCALE OF MILES
0
400
WASHINGTON
OREGON
CALIFORNIA
NEVADA
IDAHO
MONTANA
WYOMING
UTAH
ARIZONA
COLORADO
NEW MEXICO
NORTH DAKOTA
SOUTH DAKOTA
NEBRASKA
KANSAS
OKLAHOMA
TEXAS
MINNESOTA
IOWA
MISSOURI
ARKANSAS
LOUISIANA
WISCONSIN
ILLINOIS
MICHIGAN
MICHIGAN
INDIANA
OHIO
KENTUCKY
TENNESSEE
MISSISSIPPI
ALABAMA
GEORGIA
FLORIDA
SOUTH CAROLINA
NORTH CAROLINA
VIRGINIA
WEST VIRGINIA
PENNSYLVANIA
NEW YORK
NEW JERSEY
DELAWARE
MARYLAND
CONNECTICUT
RHODE ISLAND
MASSACHUSETTS
VERMONT
NEW HAMPSHIRE
MAINE
ALASKA
SCALE OF MILES
0
400
HAWAII
SCALE OF MILES
0
200

A SPRING WEATHER CHART

*WEATHER CONDITIONS:

Draw one of these symbols to represent the weather for each city.

- RAIN
- SUNNY
- SNOW
- CLOUDY

CITY	DATE	HIGH TEMPERATURE FOR THE DAY	LOW TEMPERATURE FOR THE DAY	WEATHER CONDITIONS	OTHER FACTS
1.					
2.					
3.					
4.					

What's your weather like today? ____________ Local temperature ____

Local high temperature for the day ____ Local low temperature for the day ____

SP8: Spring Tides

The term "spring tides" refers to the high and low tides that occur when the moon is in the new and full moon positions twice during a month. When the moon, earth, and sun are aligned (as in the illustrations below), these high and low tides are created. Spring tides can occur at any time of the year.

Neap tides also occur twice a month and are caused by the position of the sun and moon being at right angles. Neap tides are less dramatic low and high tides. They occur during the first and last quarter phases of the moon.

At different times of the day, the tides are higher or lower. Almanacs, newspapers, and weather stations report on these differences in height and changes in the tide. These are good resources for your students to investigate and review.

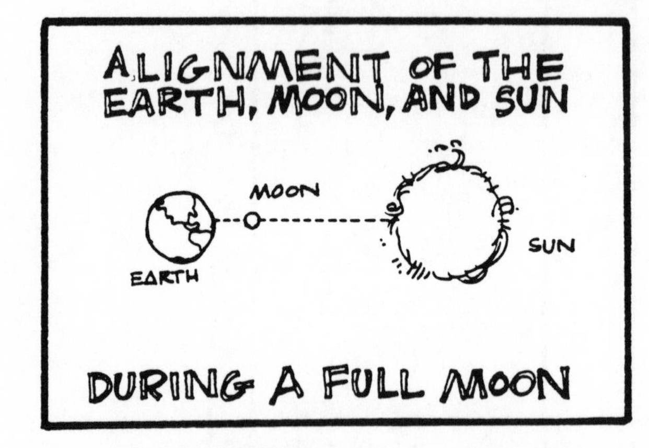

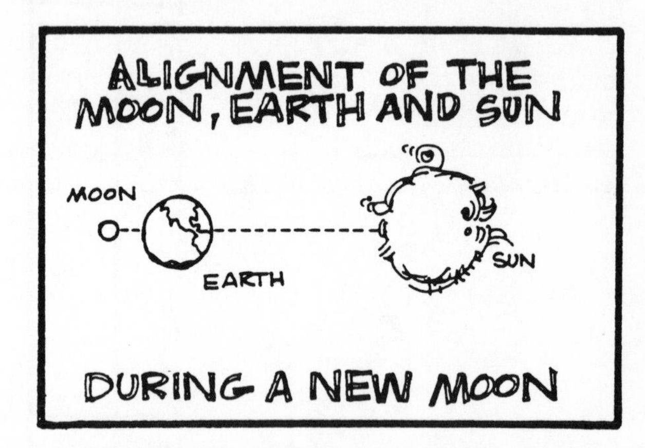

At this time, you will want to review with your students the information on the phases of the moon (see activity F10). As the moon revolves around the earth, we observe the light on the moon. When a new moon occurs, the earth and moon are on the same side of the sun. The moon is not visible at this time because we see its dark side.

However, in the full moon phase, the moon appears very full and bright. We see its lit side. The sun and moon are on different sides of the earth. It takes about two weeks for the moon to change from the new moon phase to this full moon phase. If a lunar eclipse occurs, it happens during the full phase. The sun shines on the earth and the moon is shadowed, as shown in the illustration.

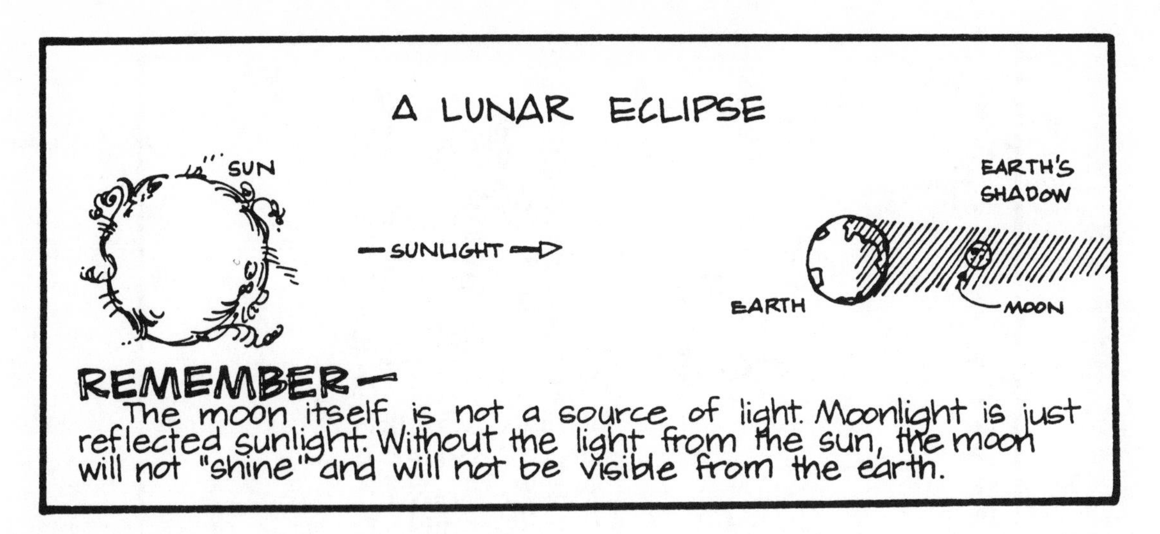

Share the concepts about the lunar eclipse with your students so they can understand "spring tides." Reproduce the two full-page illustrations and chart and give the students the opportunity to research and report on the spring tides.

Spring tides are exceptionally high and low tides affected by the alignment of the sun, moon, and earth during the full and new moon phases. The sun and moon are in the same path and exert greater gravitational pull on the earth. Gravity on the earth and the "pull" caused by the moon cause the tides to be higher. The waters rise, coming inland, and sometimes cause floods. As the earth rotates, the tides change from high to low, and vice versa.

HIGH AND LOW TIDES ARE CAUSED BY GRAVITY.

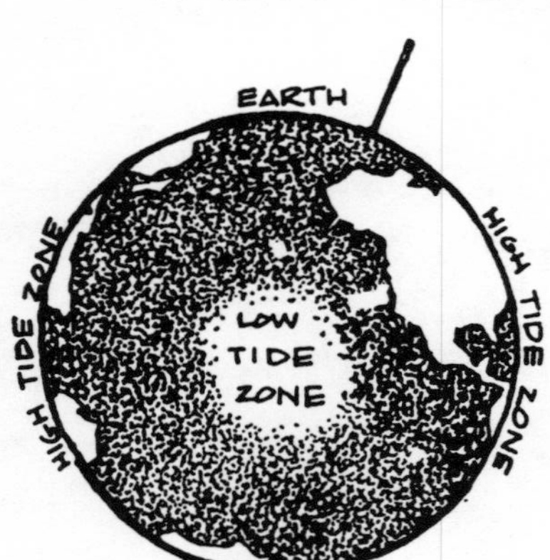

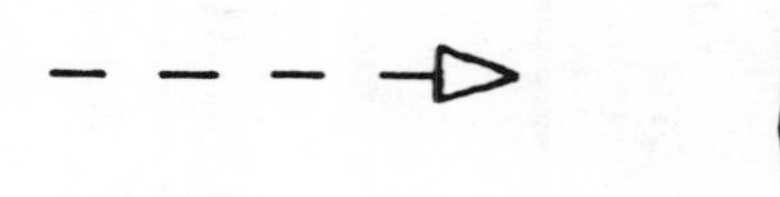

HIGH TIDES OCCUR WHERE THE EARTH IS CLOSEST TO, OR FARTHEST FROM THE MOON. WATER ON THE "NEAR" SIDE IS PULLED AWAY FROM THE EARTH BY THE MOON'S GRAVITATIONAL FORCE. MEANWHILE, THE EARTH IS DRAWN TOWARD THE MOON AND AWAY FROM THE WATER ON ITS "FAR" SIDE, CAUSING ANOTHER HIGH TIDE.

LOW TIDES OCCUR BETWEEN THESE TWO AREAS AS WATER FLOWS AWAY FROM THESE REGIONS TO THE HIGH TIDE ZONES.

BECAUSE THE EARTH ROTATES ON ITS AXIS, EVERY POINT ON THE GLOBE (EXCEPT THE POLES) MOVES THROUGH THE TWO HIGH TIDE ZONES AND AND THE TWO LOW TIDE ZONES EVERY DAY. THAT IS WHY WE HAVE TWO DAILY HIGH TIDES AND TWO DAILY LOW TIDES.

THE SUN ALWAYS ILLUMINATES THE HALF OF THE MOON THAT IS FACING IT.

WE SEE THE HALF OF THE MOON THAT FACES THE EARTH.

MOON

WE SEE THIS SIDE OF THE MOON FROM THE EARTH

THE APPEARANCE OF THE MOON IN THE NIGHT SKY DEPENDS UPON HOW THE HALF OF THE MOON THAT FACES THE EARTH IS ILLUMINATED BY THE SUN

(LAST QUARTER)
The part of the moon we see is half lit and half in shadow.

(NEW MOON)
The part of the moon facing the earth is in shadow

EARTH

The part of the moon we see is fully lit.
(FULL MOON)

The part of the moon we see is half lit and half in shadow.
(FIRST QUARTER)

Name ________________________ Date ________________________

Spring Tides Chart

Source of information (newspaper, television, etc.):

____

Time of morning high: ____________

Time of morning low: ____________

Time of evening high: ____________

Time of evening low: ____________

Phase of moon: ____________

Name of city or dock: ________________________

Review illustrations of the moon, earth, and sun. Then draw a picture below of the alignment of the earth, sun, and moon at high tide.

Name ______________________________

SP9: Averaging Spring Temperatures

Here is a chart to help you gather information on the weather in your area or in another part of the country. You may want to do one local report and one from a different city. First, record the temperature highs and lows on the chart for seven days. Then add the seven high temperatures (get permission to use a calculator). Add the seven low temperatures. Record both sums on the back of this sheet. Divide each sum by 7, representing the seven entries you made. This answer (the "quotient" in division) will give you the average WEEKLY HIGH and the average WEEKLY LOW. Record them on the back of this sheet also. If you are learning long division in math, or you already know it, then show all your math computations on the back of this sheet. If not, and you have permission to use a calculator, then enter your answers only.

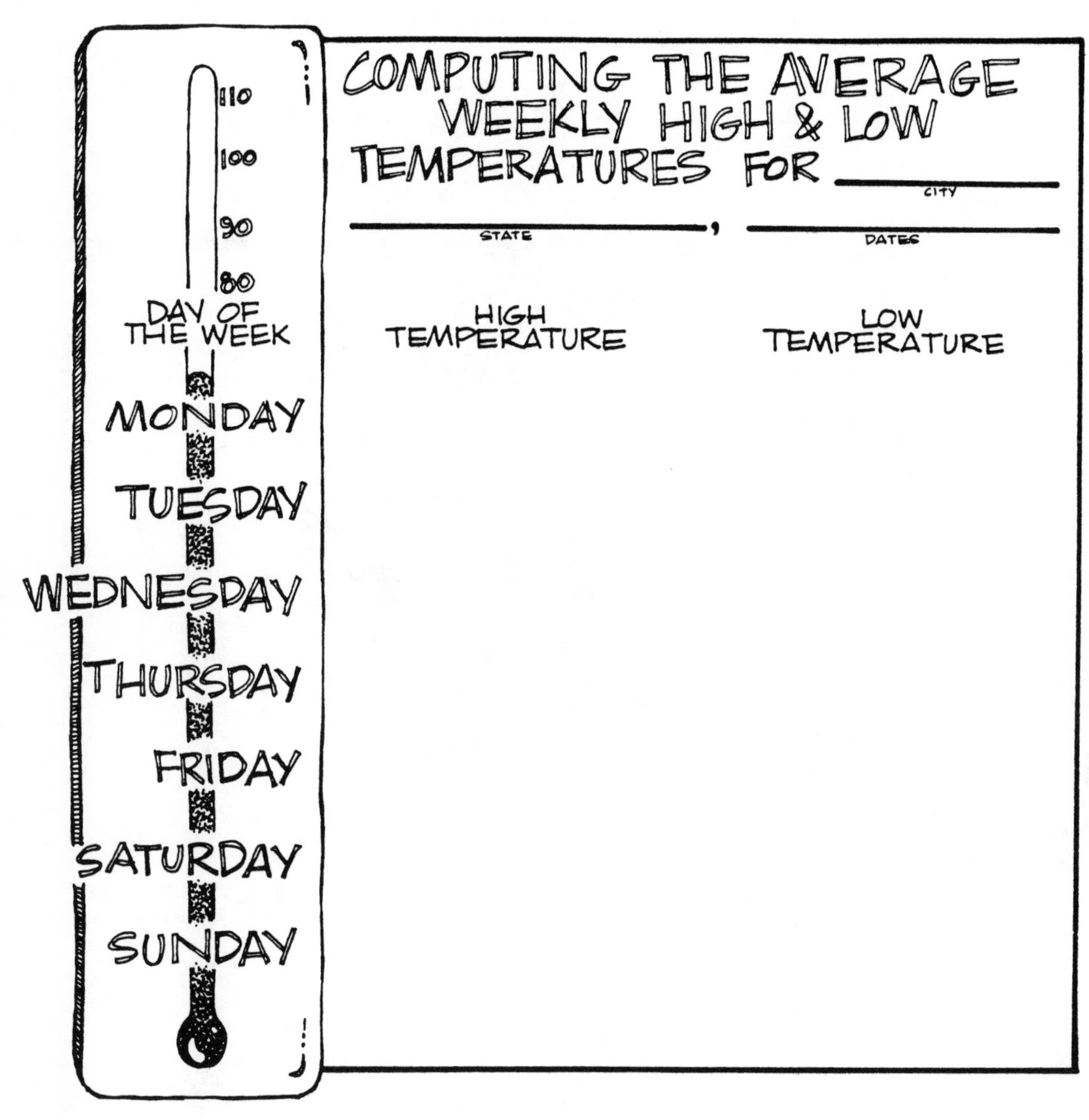

SP10: Daylight Saving Time

Daylight Saving Time is defined as an extra hour of daylight added to Standard Time. Standard Time is a zone time, referring to an area of local time defined by meridians of longitude (an angular measure of reference on a globe). There are 24 different meridians of longitude (each belt measures 15° of longitude), and in the United States, there are four basic time zones: Eastern, Central, Mountain, and Pacific.

In the spring, we change our clocks. Daylight Saving Time begins around April 27. It begins at 2:00 a.m. Eastern Time. It lasts until the fall, around October 26–29.

A few places have two time zones because they are on the boundary line between time zones. If you look at a globe and move from east to west, subtract one hour from each zone to determine the time. For example, if it is 3:00 p.m. Eastern Time, then it is 2:00 p.m. Central, 1:00 p.m. Mountain, and 12:00 noon Pacific.

Enlarge the following illustration on an opaque projector and reproduce the diary for each student. The students can use the displayed map while completing the diary.

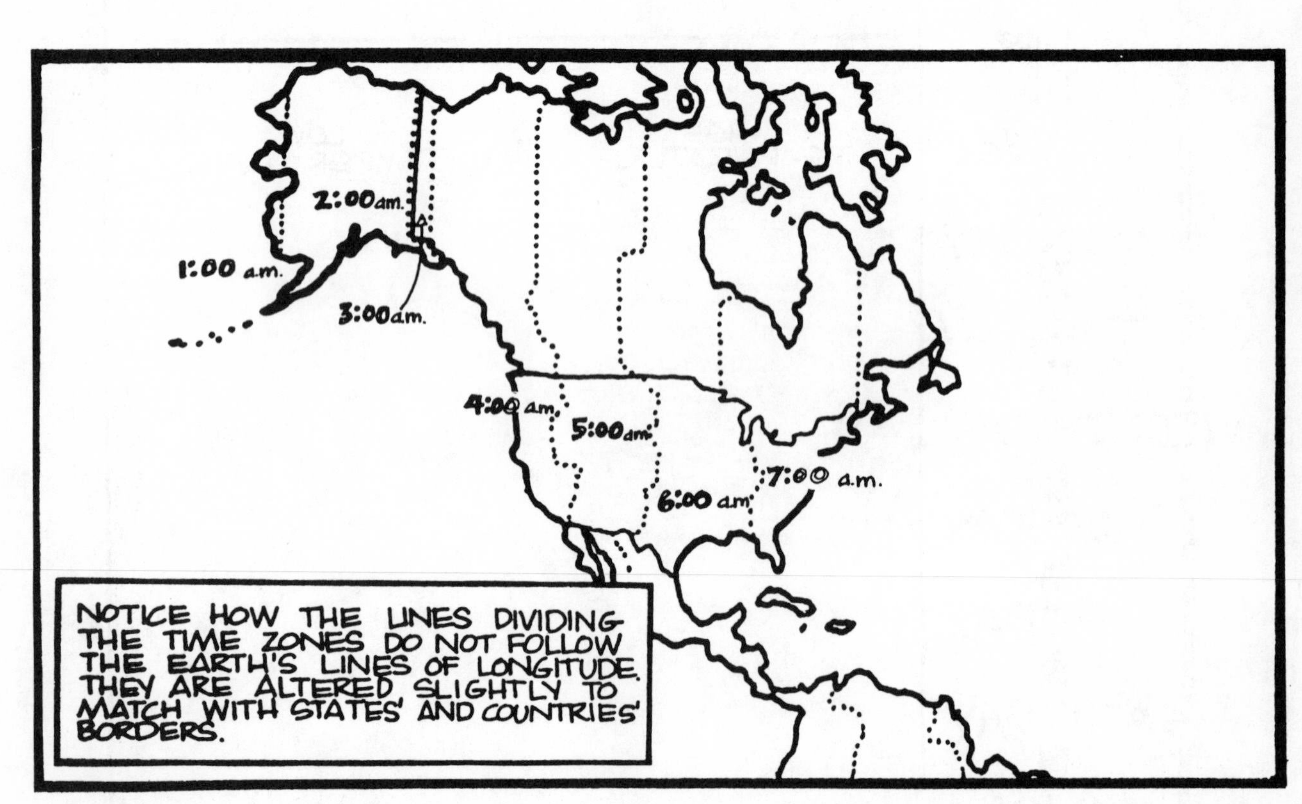

A

DAYLIGHT SAVING TIME DIARY
AN ACTIVITY RECORD

1. ON WHAT DAY DOES DAYLIGHT SAVING TIME BEGIN THIS SPRING?

2. WHEN WILL IT END?

3. WHEN DID YOU TURN YOUR CLOCKS AHEAD?

4. WHAT IS YOUR TIME ZONE?

5. THE TIME ZONE DIRECTLY TO YOUR EAST IS __________; TO YOUR WEST IS __________.

6. IF IT IS 6:00 P.M. CENTRAL STANDARD TIME, WHAT TIME IS IT MOUNTAIN TIME?

7. NAME ONE STATE IN EACH TIME ZONE.

EASTERN __________

CENTRAL __________

MOUNTAIN __________

PACIFIC __________

8. LOOK ON A GLOBE AND FIND THE LINES OF LONGITUDE THAT DIVIDE THE UNITED STATES' TIME ZONES. WHAT ARE THEY?

Diary

Name ______________________________ Date ________________________

SP11: What Is the Vernal Equinox?

The Vernal Equinox ushers in the first day of spring. On this day, night and day are of equal lengths—just like the Autumnal Equinox.

Find out when the Vernal Equinox begins and write the date here: ______________________. Look in the newspaper for articles on this event.

Using the second part of this activity to write your answers, try to estimate the time of sunrise and sunset on this day. Then use the local weather and television news reports to confirm this information.

Name ______________________ Date ______________________

SP11: What Is the Vernal Equinox? (continued)

My estimated time of sunrise ________

My estimated time of sunset ________

LOCAL TELEVISION WEATHER REPORT

Time of sunrise on the Equinox ______

Time of sunset on the Equinox ______

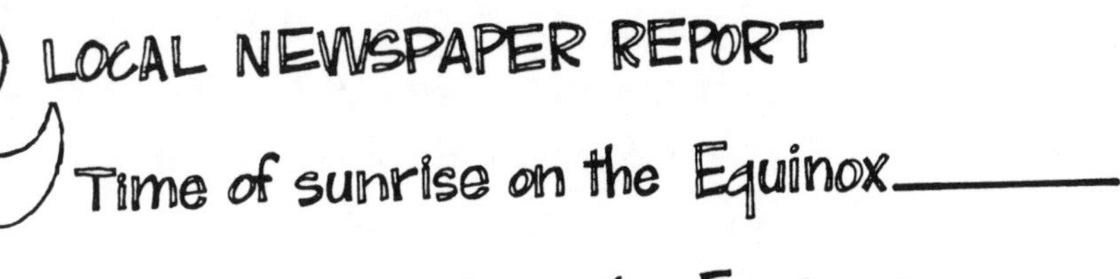

LOCAL NEWSPAPER REPORT

Time of sunrise on the Equinox ______

Time of sunset on the Equinox ______

SP12: The Spring Sky—Looking for Leo

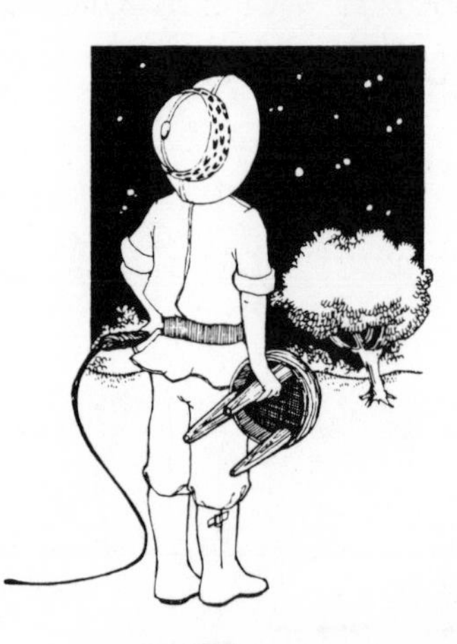

The spring sky contains many beautiful constellations, Leo being one of them. Review the major constellations for viewing, including Centaurus, Boötes (Bō-ōt-ēz), Virgo, and Leo.

Similar to F12 and W12, have the students make constellation viewers with either oatmeal boxes or styrofoam cups to view Leo. Two Leo patterns are given—one is for the oatmeal box, the other is for the styrofoam cup.

This is a good time to introduce students to the temperatures of stars and their distance in light years.

HOW BIG IS A LIGHT YEAR?

186,000 MILES PER SECOND (SPEED OF LIGHT)
X 60 NUMBER OF SECONDS IN A MINUTE
11160000
X 60 NUMBER OF MINUTES IN AN HOUR
669600000
X 24 NUMBER OF HOURS IN A DAY
16070400000
X 365 NUMBER OF DAYS IN A YEAR
5,865,696,000,000 NUMBER OF MILES LIGHT TRAVELS IN A YEAR (ONE "LIGHT YEAR")

THIS NUMBER IS READ "FIVE TRILLION, EIGHT HUNDRED SIXTY-FIVE BILLION, SIX HUNDRED NINETY-SIX MILLION."

WHAT DOES THIS MEAN?

Light can make the trip from the sun to the earth* 61,614½ times a year, or 169 times a day, or in just under 10 minutes.

*THE SUN IS OVER 95,000,000 MILES FROM THE EARTH.

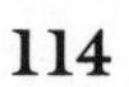

Centauri is one of the brightest stars and is in the constellation Centaurus. It is a yellow star and is the third brightest star in the sky, with a temperature of 10,000°F. It is 4.3 light years away.

Arcturus, the sixth brightest star, is an orange star found in Boötes. It is 32 light years away. Virgo contains Spica, a blue-white star with a temperature of about 40,000°F.

Reproduce "The Spring Sky" activity sheet for your students, as well as the information on the four major spring constellations.

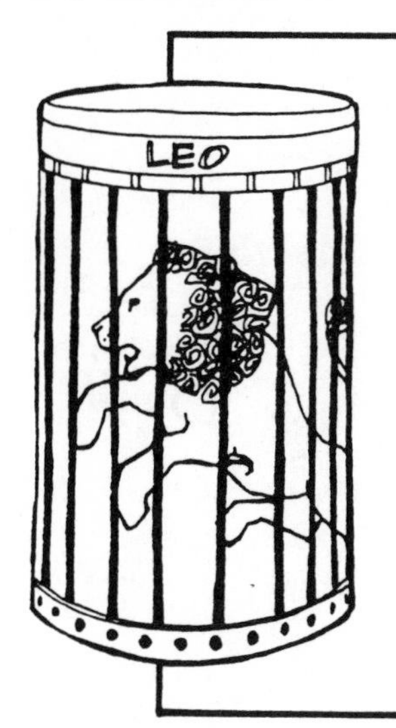

Name ______________________________ Date ______________________________

Patterns for Leo

Use this pattern if you are using an oatmeal box:

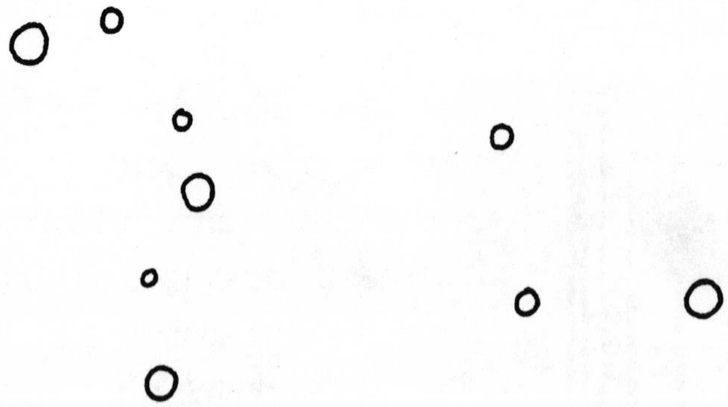

Use this pattern if you are using a styrofoam cup:

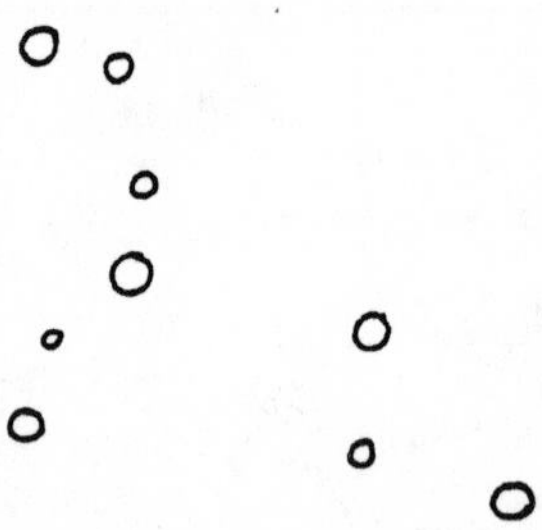

The Spring Sky—Information Sheet

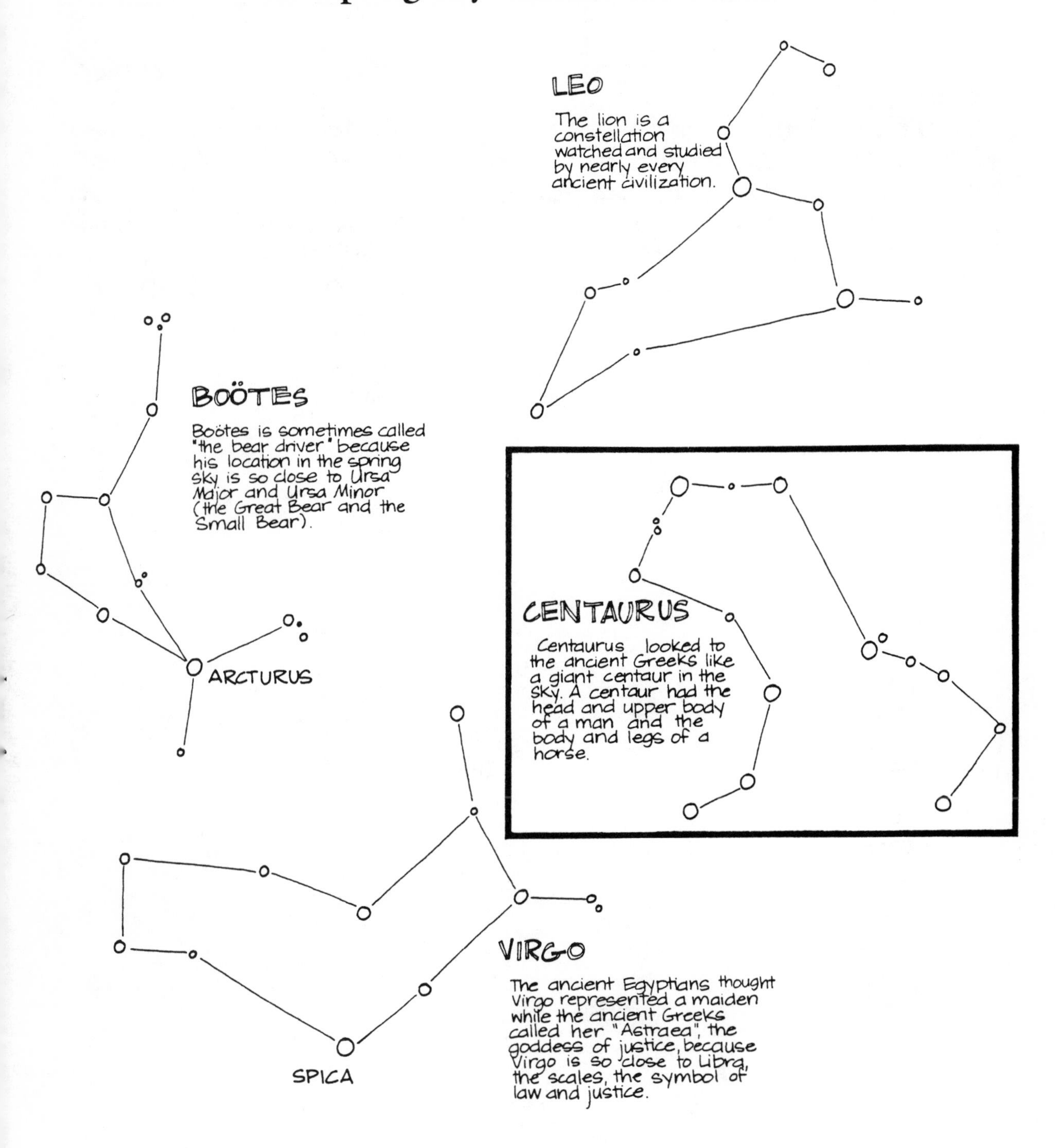

Name______________________ Date______________________

When the Greeks, Egyptians, and other ancient peoples looked at the night sky, they saw animals, monsters, and characters from their myths drawn in the patterns of the stars. Look at the four constellations on this page. Can you tell how those ancient stargazers might possibly have seen a lion, a herdsman, a centaur, or a maiden in the constellation Leo, Boötes, Centaurus, or Virgo? Now draw a picture of each of these mythical figures over the pattern of its constellation.

LEO
THE LION

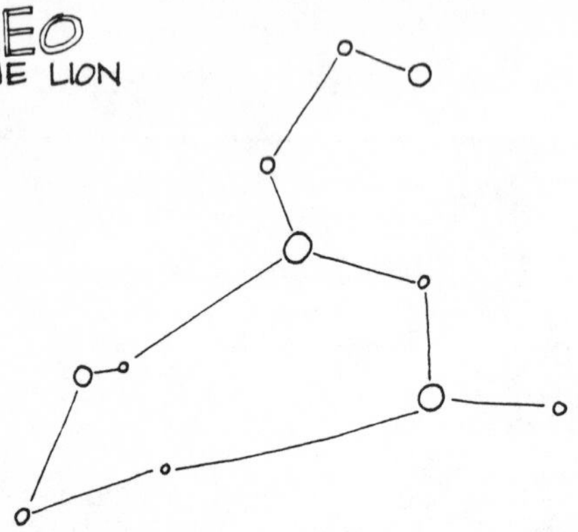

BOÖTES
THE HERDSMAN

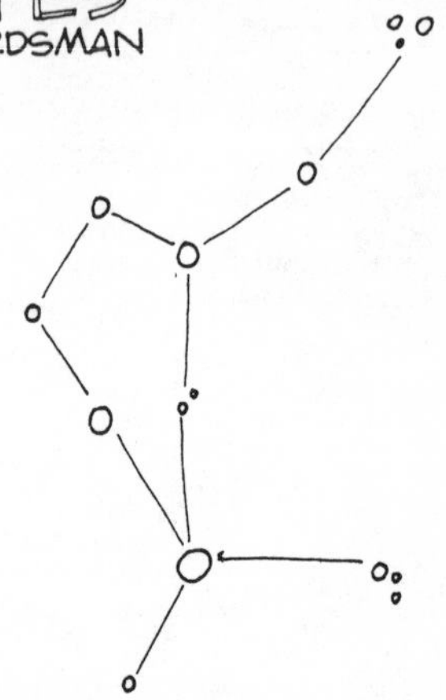

CENTAURUS
THE CENTAUR

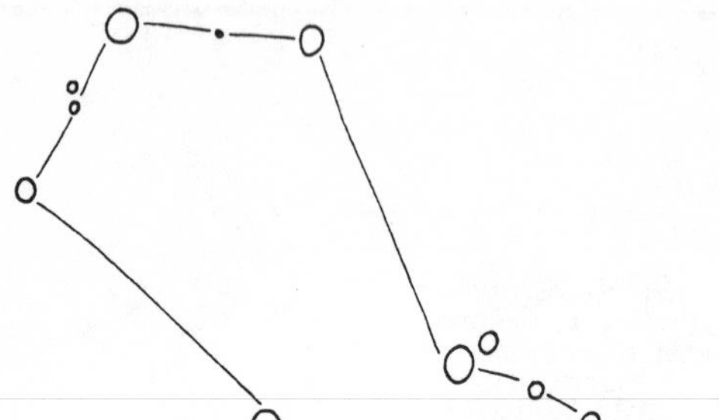

VIRGO
THE MAIDEN

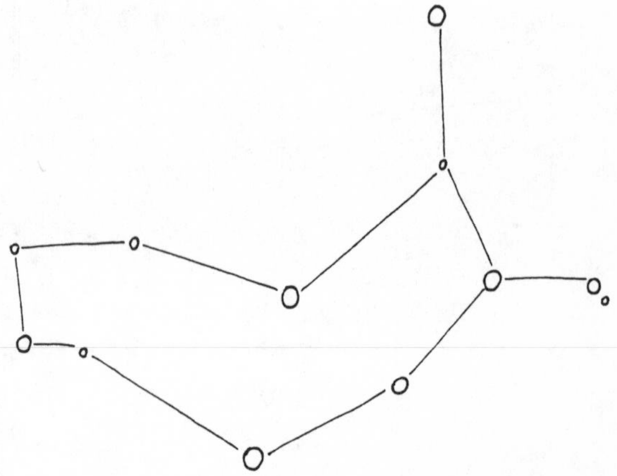

THE SPRING SKY

SPRING
1. enrichment
2. activities

Spring Cookery—Cream of Broccoli Soup

Spring is a good time of the year to plant seeds with your class, or start a school or home vegetable garden. The "Cool and Warm-Weather Crops" activity sheet asks the children to sort and name cool- and warm-weather crops. If you plant some of these with your class, students will experience the fun and enrichment of seeing their crops grow.

Below is a recipe for "Cream of Broccoli Soup." Since broccoli is a cool-weather crop, this is a good time of the year to make some with your class. One cup of broccoli has 45 calories, 5 grams of protein, and 8 grams of carbohydrates. It also has 2 grams of fiber, so you may want to discuss its nutritional value with your class. (Special thanks to Doreen Krohner of Avon, CT, owner of Doreen's Delights, for this recipe.)

EQUIPMENT/INGREDIENTS:

1 pound of fresh broccoli
1 small onion
2 medium potatoes
1 chicken bouillon cube
3 cups of water
1 cup of milk

Knife (CAUTION: To be used only by an adult)
Saucepan
Measuring cup
Electric blender
Pot holders
Heat source (CAUTION: To be used only under adult supervision)

DIRECTIONS:

1. Wash the broccoli and shake dry. Remove the leaves.
2. Peel the stalks with the knife. Discard the outer peeling.
3. Cut the stalks and head into bite-size pieces. Place pieces into a saucepan.
4. Peel and rinse the potatoes and cut into quarters. Peel and rinse the onion and cut into quarters. Add these quarters to the saucepan along with water and bouillon.
5. Cook vegetables until tender over medium to medium-low heat.
6. Cool slightly and put all into a blender. Blend until creamy.
7. Add the milk and mix well.
8. Return everything to the saucepan and reheat until hot. (For thinner soup, add more milk.) This recipe serves four.

Name____________________ Date____________________

COOL- AND WARM-WEATHER CROPS

From the pictures, list the 14 COOL-WEATHER CROPS.

1. ____________
2. ____________
3. ____________
4. ____________
5. ____________
6. ____________
7. ____________
8. ____________
9. ____________
10. ____________
11. ____________
12. ____________
13. ____________
14. ____________

List the 10 WARM-WEATHER CROPS.

1. ____________
2. ____________
3. ____________
4. ____________
5. ____________
6. ____________
7. ____________
8. ____________
9. ____________
10. ____________

SOMETHING TO THINK ABOUT: From what parts of the plant do the cool-weather crops come? How about the warm-weather crops?

Using Solar Energy to Cook

This activity involves using a fresnel lens (magnifying lens) to cook and focus the sun's rays. Since working with a fresnel lens requires adult supervision, this activity should be done as a teacher demonstration. Caution is advised because students should not look directly at the focal point of the lens nor stare directly at the sun.

As you experiment with the lens, you will learn control of the lens. Use caution with your fingers, hands, and clothing as any lens in direct sunlight can ignite or burn them. (Special thanks to Jonathan Craig, Director of Ecology at Talcott Mountain Science Center in Avon, CT for this activity.)

EQUIPMENT:

Fresnel lens (CAUTION: This magnifying lens to be used with extreme care)

Large tin can or cooking pot

Pot holder

Can of soup or water

Can opener

Wooden spoon

Sunlight

DIRECTIONS:

1. Open the can of soup and pour the soup (or water) into the tin can or pot.
2. Take the fresnel lens and angle it so the focal point diffuses the sun's rays over the cooking surface of the soup or water. This will vary depending on the size of the container.
3. Remove the lens and stir the soup to distribute the heat.
4. Continue heating until the soup is warmed.
5. How long did it take? You may want to try this on different days and at different times. A good time might be at noon during the month of May!

Name ____________________ Date ____________________

Spring Science Stories, Songs, & Poetry

DECODE ME!

Decode the names of these famous constellations:

Write four more here

____

____

____

____

and have a friend decode them.

THE CODE

A	B	C	D	E	F	G	H	I	J	K	L	M
N	O	P	Q	R	S	T	U	V	W	X	Y	Z

Name________________________ Date ________________________

SPRING SCIENCE CRAFTS: SPELLING ART

Create a picture and spell its parts. Try any of these:

1) Parts of a flower
2) Parts of a frog
3) Parts of an insect
4) Parts of a tree

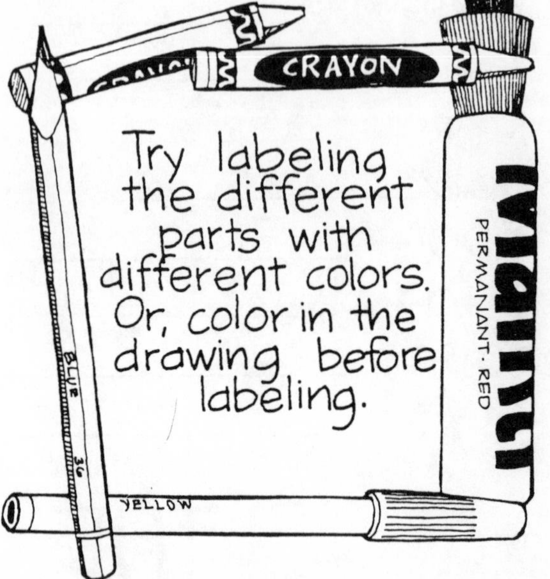

Here's an example:

Name ______________________ Date ____________________

Spring Science Stories, Songs, & Poetry

WRITING DIAMANTES

Here is an easy way to write a poem about the beauty that surrounds us at Springtime. Just fill in the blanks!

- On the first line, write down the name of a person, place, or thing (NOUN). → BUD
- Under that write two words that describe that noun (ADJECTIVES). → SMALL SHELTERED
- Then write three action words (VERBS). → REACHES OPENS UNFOLDS
- On line four write two words that describe the verbs you've just written (ADVERBS). → SLOWLY BEAUTIFULLY
- Finally write another noun with the same meaning or DEFINITION as the word you wrote on line one. → FLOWER

*Try to make all your words relate to a THEME or topic. The theme of this poem is Spring Flowers.

NOW YOU TRY ONE

______
NOUN

______ ______
ADJECTIVE ADJECTIVE

______ ______ ______
VERB VERB VERB

______ ______
ADVERB ADVERB

______
NOUN

Name ____________________ Date ______________

SPRING SCIENCE STORIES, SONGS & POETRY

"SIGNS OF SPRING"

Music by: Patricia McCamish Donohoe   Lyrics by: Julia Moutran

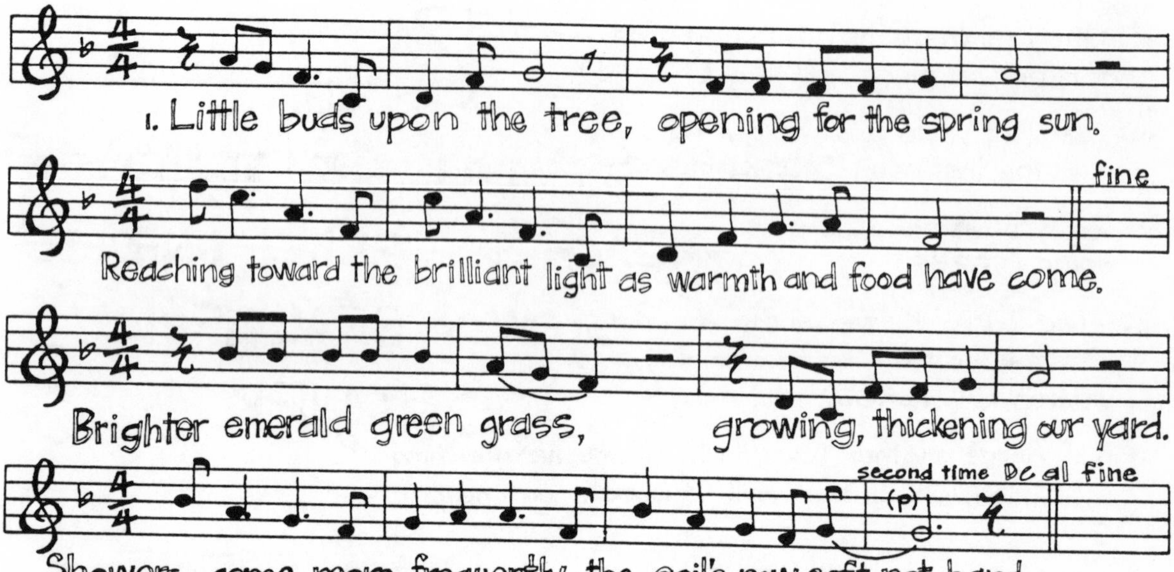

1. Little buds upon the tree, opening for the spring sun.

fine

Reaching toward the brilliant light as warmth and food have come.

Brighter emerald green grass, growing, thickening our yard.

second time DC al fine

(p)

Showers come more frequently; the soil's now soft, not hard.

2. Baby birds, insect larvae—
Stages of new life everywhere.
Eggs hatching, people catching
Spring fever's in the air.

3. Signs of spring at last have come;
Winter's long rest has passed.
Three months of beauty and growth
Have come to us at last!

With Guitar Accompaniment

SPRING METAMORPHOSIS

Spring is a time of change and new life. Children will find tadpoles in the brooks, birds returning north, and insects emerging from cocoons. This is a time of metamorphosis.

The following activity pages have crossword puzzles and a matching sheet on insects, including butterflies, moths, grasshoppers, and other types. A good book for reference as well as for classroom use is *Collecting Bugs and Things* (published by Price, Stern, Sloan, Inc.). See page 180 for ordering information.

Field guides are also very valuable for identification and collecting. Your students will need to research grasshoppers, butterflies, and moths to complete the activity sheets and the two crossword puzzles. You may want them to explore the characteristics of grasshoppers and crickets—the nymph stages are found in the spring, as are baby frogs in the tadpole stage. If your students collect any of these specimens and bring them into class, be sure to have them set up homes similar to the animal's natural environment and provide food for it (if needed). Encourage them to find out where the adult lays the eggs over the winter and how they survive to be born in the spring!

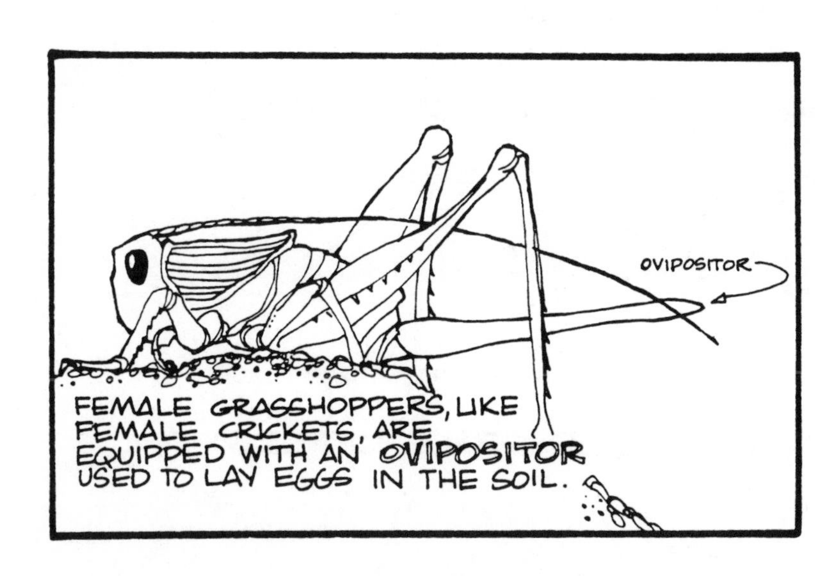

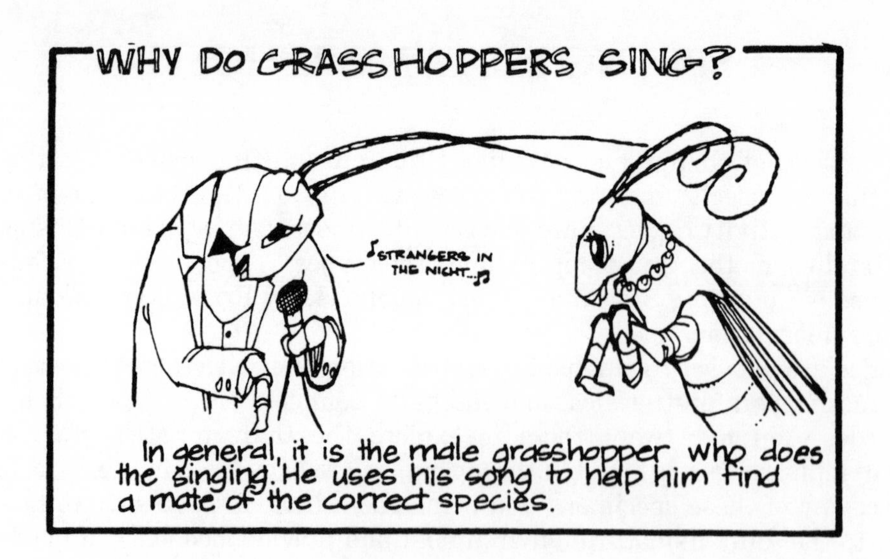

Actually, grasshoppers "sing" by rubbing two body parts together.

LONG-HORNED GRASSHOPPERS

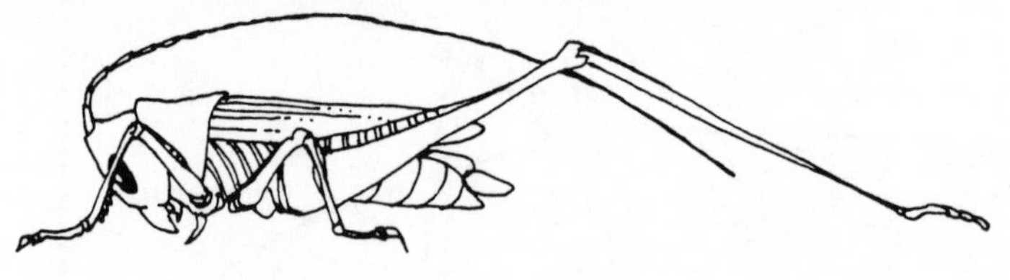

rub their wings together.

SLANT-FACED GRASSHOPPERS

rub their hind legs and their wings.

BAND-WINGED GRASSHOPPERS

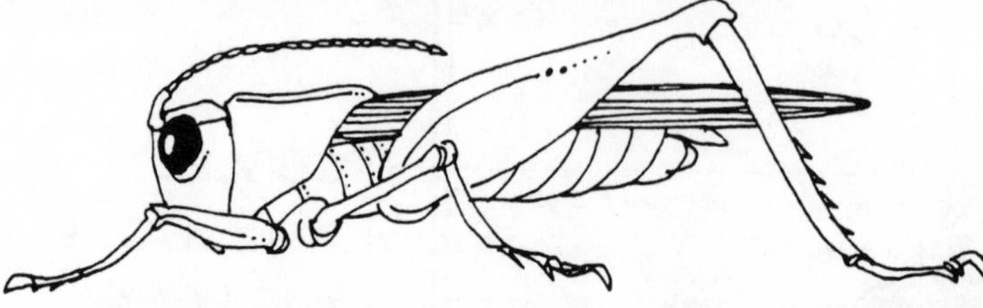

clap their wings together as they fly.

Characteristics of Tadpole	Characteristics of Frog
1. Has Gills	1. Has Lungs
2. Has Tail & Mouth Suckers	2. Has Legs & Arms & Teeth
3. Swims Only – Likes Water	3. Swims & Jumps – Likes Land & Water
4. Has Fins	4. Has Fingers & Webbed Feet
5. Eats Algae	5. Eats Insects: Flies & Moths

Name ______________________ Date ______________________

MATCHING: FUN WITH INSECTS

Scientists who study insects are called **ENTOMOLOGISTS**. To make their job a little easier, they separate the different types of insects into groups, or "classes." The common names for five types of insects are listed in the column in the left; the names of their scientific classifications are in the right hand column. See if you can match the common name to the scientific name by drawing a line connecting the two of them.

COMMON NAME	SCIENTIFIC CLASSIFICATION
Flies	Hymenoptera
Beetles	Lepidoptera
Dragonflies	Odonata
Bees, Wasps, Ants	Diptera
Moths, Butterflies	Coleoptera

Now write the name of the scientific classification under the picture of each insect.

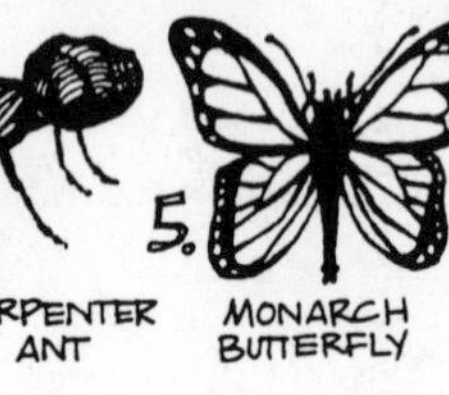

_______ _______ _______ _______ _______ _______ _______ _______

ANSWERS:

Matching: Flies (Diptera), Beetles (Coleoptera), Dragonflies (Odonata), Bees, Wasps, Ants (Hymenoptera), Moths, Butterflies (Lepidoptera)

Naming: 1. Hymenoptera 2. Coleoptera 3. Odonata 4. Hymenoptera 5. Lepidoptera 6. Diptera 7. Coleoptera 8. Lepidoptera

Name________________________

Date________________________

SPRING CROSSWORD

PUZZLE I

ACROSS

2. To shed the skin or exoskeleton in the process of growth
3. Butterfly that flies South for the winter and lays eggs on milkweed
5. The second stage of a grasshopper's metamorphosis
8. The immature, wingless, feeding stage of an insect
9. Number of body parts of most insects
10. Season when many insects lay eggs
11. Season when most insects are in the larva form.

DOWN

1. The silky cover spun by the larva that serves as protection while the insect is in the pupal stage.
4. Name of the butterfly pupa
6. Season when butterflies and moths emerge from cocoons
7. Larval form of a butterfly or moth

Name ________________________ Date ____________________

SPRING CROSSWORD PUZZLE II

DOWN

1. A nutritious cool-weather green vegetable
4. Cluster of modified leaves near the base of the petals of a flower that helps protect the flower before it opens
6. Preliminary or "baby" stage of a grasshopper
7. Name of scientific order or classification of the grasshopper

ACROSS

2. Word meaning "equal night"
3. Type of magnifying lens used to magnify or focus the sun's rays
5. Force that causes high and low tides
8. Substance used to make soil more fertile

The Summer Season

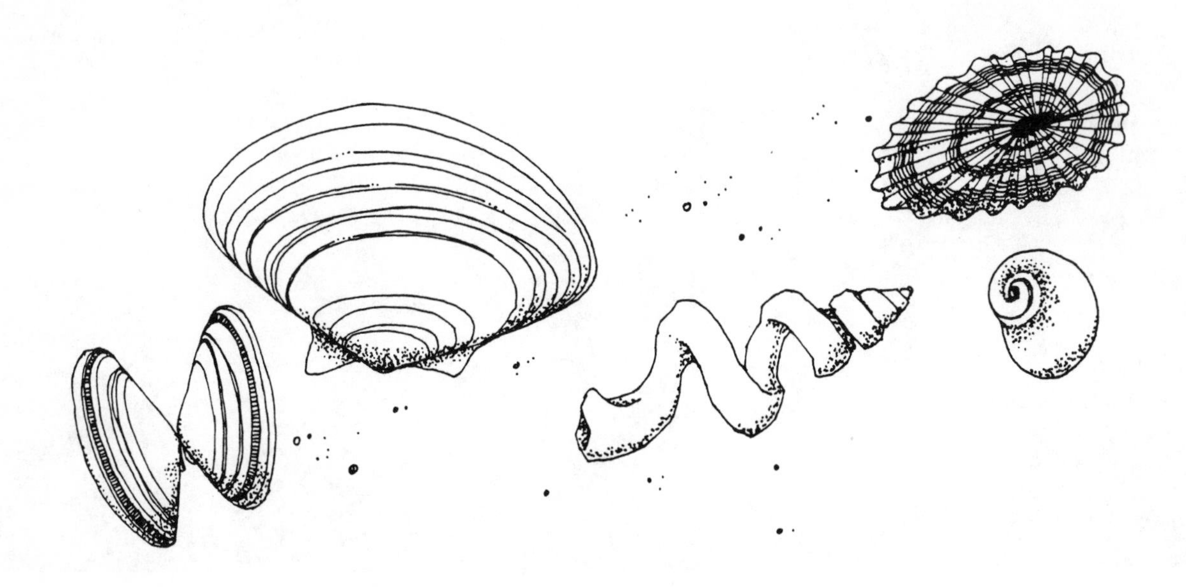

TEACHER'S GUIDE FOR SUMMER

This is the fourth and final section of the book, representing activities that can be done entirely by the student at home or in school. They can be done at camp or in summer school, or as a gift to the students for the summer vacation! If you live in a warmer climate, these activities can be done at any time of the year.

You may want to do a combination of the above. Peruse and select some activities from the twelve lessons to be done in school and, perhaps, send all of the nine enrichment lessons home. Or, select a few activities for the weeks of school during June, and send home a package for the vacation months. All of the activities are shorter and more self-directed; only a few need adult supervision.

Many of the activities will give the students further opportunities to apply their concepts and skills from the school year in another learning environment. Also, if your students do not attend summer camp and could benefit by some extra educational direction and opportunity during the summer, this "gift" is truly one that will help them learn and have fun at the same time.

If you plan to send the activities home as a gift to your students, you will want to reproduce the sheets and review some or all of them with your students. Motivation and interest will be greater if you spend some time orienting them to the activities. Have students make folders or science notebook covers (with the full-page illustration on page 134). Let the students color them and decorate them. Bind the notebooks in school with paper fasteners or colorful yarn. The students will be proud to take the notebook home and show it to their families.

Give your students copies of the Science Season Certificate. Sign it ahead of time and enclose it as the last page of the science notebook. Then the parent can sign and award it to the child after the activities are complete. If you want your students to do only some of the activities, be sure to indicate that to the parent on the Parent Letter. Send the letter along with the student science notebook.

In preparation for this unit, you may want to show some of the correlated science videos recommended in the Appendix.

You may also want to make a bulletin board or showcase on different types of summer clouds. You could also make a sample Cloud Mobile for the students to see.

Talk to the students about the changes in the earth's journey around the sun and about energy from the sun. Using the sun's energy, you may want to make a

Parabolic Hot Dog Cooker to show students what the final product looks like and talk to them about parabolas.

IMPORTANT: Review the activities that require adult supervision. There are only a few that require adult or teacher supervision, such as cutting a potato or vegetable for planting or printing.

Sing the science song, "My Favorite Season," with your students so they will be familiar with the music. Talk about the helpful ways bees and butterflies pollinate trees and plants and their importance in the garden.

When viewing the full-page illustration, be sure to discuss these observations:

- Tadpoles have become frogs.
- Animals are out and about.
- Bulbs are past blooming.
- The garden is growing.
- Blossoms have become ripening apples on the trees.
- The sheep have been sheared.
- Shadows are short.
- The brook is thinner due to summer heat.
- Dandelions grow among the grass as food for certain animals, such as groundhogs.
- Thunderheads are typical of summer rain clouds.
- The children's clothing reflects changes in season temperatures.

It is hoped that you have enjoyed using *Elementary Science Activities for All Seasons*, and that you will continue to use the book again and again, season after season!

Dear Parents,

The Summer Solstice announces the arrival of summer! As part of our science program, your child and I have prepared a special Summer Science Notebook that can be completed during the summer vacation.

Inside this book are science activities and enrichment activities about the summer season. We have learned about other seasons this year, and this will give your child a chance to enrich his or her summer vacation with some fun and interesting science investigations.

Review the notebook and help your child select the ones he or she wants to complete. The materials are simple and few. It would be helpful if you read the activities and help provide the materials needed for each activity.

At the end of the notebook or selected activities, your child will receive the special Science Season Certificate. This certificate can be signed by you when you feel your child has earned it.

I hope that you and your child have a wonderful summer vacation and enjoy completing this Summer Science Notebook. I have enjoyed having your child in my class this year!

Sincerely,

____

Teacher

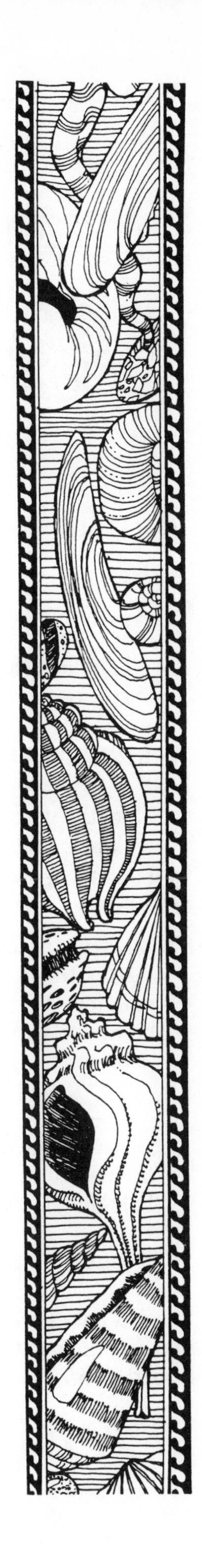

Name ________________________________ Date ____________________________

S1: What and When Is Summer?

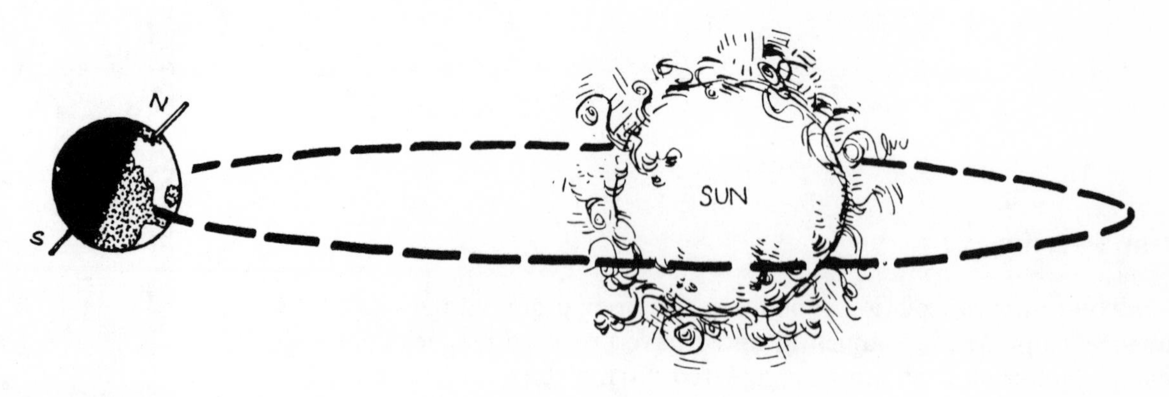

EQUIPMENT:

Ball
Flashlight
2 small pieces of paper
Pencil
Tape
String
Scissors

DIRECTIONS:

1. Take the papers and write "North Pole" on one and "South Pole" on the other. Cut them out as labels and tape to the ball on each end.
2. Cut the string to fit around the ball and tape to the middle of the ball to represent the equator.
3. Look at the illustration at the top of this sheet. It represents the position of the earth during the summer in its journey around the sun. Look at the position of the poles and the equator. When the summer begins on the Summer Solstice, the longest day of the year occurs. "Solstice" means the sun stands still. Find out when that day is this year and record it here:

   ____

4. Go into a darkened room with the ball (model of the earth) and the flashlight. Illuminate the ball with the light and observe the position of the earth. Shine the light on the North Pole directly. What do you observe? What happens to the Southern Hemisphere?

CONCLUSIONS:

What can you conclude about the Southern Hemisphere and the Northern Hemisphere during the months of June, July, and August? Read and find out about the differences in their weather and seasons.

Name ______________________________ Date ______________________

S2: Examining Cloud Types

Summer is a great time to observe differences in clouds. Look at their shapes and their colors, and see if you can tell the differences. Look in an encyclopedia or a book on clouds from your library to find out about the differences in clouds. There are more than ten different kinds!

Most clouds are cumulus or stratus types. Cumulus are puffy clouds, and stratus are layered. Clouds are formed from water vapor in the earth's atmosphere. The condensation that is formed from water vapor being cooled creates clouds.

Use the chart to examine clouds for eight different days. Write in the date, the name of the cloud type you observe in the morning and afternoon, and the weather at the time. If you want to make a Cloud Mobile, look at the enrichment activity, "Making Cloud Mobiles."

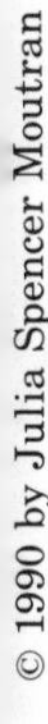

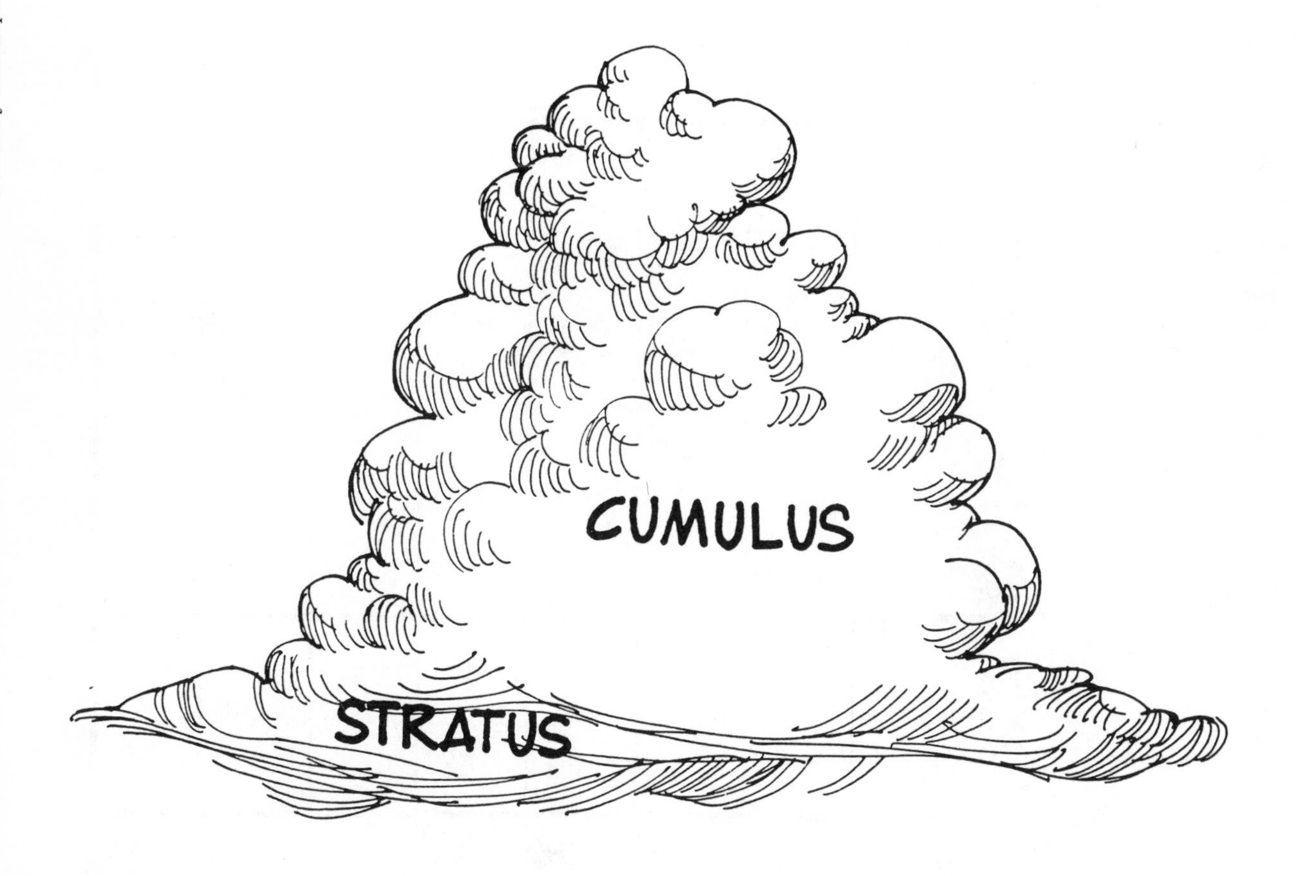

Date	Clouds in the Morning	Weather	Clouds in the Afternoon
1.			
2.			
3.			
4.			
5.			
6.			
7.			
8.			

Name ______________________________ Date ________________________

S3: Examining Shadows at Noon

Use "My Shadow Chart" to record the time and length of your shadows. Do you think the sun's shadows are longer or shorter in the summer than in the winter? See if you can find out. Here is a hint: The sun is higher in the sky in the summer than in the winter.

Measure your shadow at three different times of the day—morning, noon, and late afternoon. Compare the differences.

EQUIPMENT:

Stake to mark your shadow
Yardstick or piece of rope
"My Shadow Chart"
Pencil
A friend
Sunshine

DIRECTIONS:

1. Go outdoors three different times to measure your shadow on a sunny day. Have your friend measure the length of your shadow using the yardstick (or rope) and stake. Record the information on the chart.
2. Compare the shadow measurements for the three times of the day.
3. You may want to do this on the Summer Solstice and during different times of the summer and compare your findings.

CONCLUSIONS:

When was your shadow the longest? Why?

Name ________________________ Date ____________________

S4: Graphing Summer Temperatures

For this activity, you will select two cities and graph and record their temperatures for one week. Get your temperature readings by reading the newspaper, watching the weather report on television, or writing a pen pal and exchanging information. Maybe your grandparent or a friend living in another city will keep a record of the weather in his or her town and exchange the information with you.

Using the graph, put a dot to represent the temperature for your city or town. Connect the dots at the end of the week, drawing a line from one dot to the next. Graph the other city's weather in a contrasting color. Compare the differences. Print the cities' names on the chart. If you know how to average temperatures, you may want to average the weekly HIGH TEMPERATURES. When graphing, you will be recording the high temperatures for each day.

Cities' Names
Dates
DEGREES FARENHEIT
110°
100°
90°
80°
70°
60°
50°
40°
30°
MONDAY
TUESDAY
WEDNESDAY
THURSDAY
FRIDAY
SATURDAY
SUNDAY

Name ________________________________ Date ________________________

S5: Measuring Relative Humidity

Relative humidity is a measurement expressed as a percentage, an index of the condition of the environment. In part, it measures the water content of air. Scientists and meteorologists use formulas (mathematical equations) to calculate the relative humidity.

Relative humidity can change during the course of the day. An air temperature of 70 degrees, for example, might have a relative humidity of 58 percent. The higher the percentage (as it approaches 90, for example), the higher the humidity.

When it is humid outside, how do you feel? Many people feel uncomfortable on very humid days.

A weather instrument called a "hygrometer" is designed to measure relative humidity. Sometimes a high reading of relative humidity indicates rain may be coming.

When thinking about a hygrometer, think in terms of wetting your hair. When your hair (especially if you have bangs) is wet, it seems longer. As it dries, the hair shrinks. The water content is greater in the wet hair, and less in the dry hair.

This lesson has two activities: one experiment and one chart. The experiment will measure the shrinkage in wet and dry hair, similar to the index used in a hygrometer.

EQUIPMENT:

4 different hairs of varying length, color, and texture

Sponge

Water

Ruler

Paper

Pencil

Tape

Clock or watch with a minute hand

DIRECTIONS:

1. Measure each strand of hair while dry. Tape the hair at both ends and straighten each one to measure. Record the length, color, and texture (fine, coarse, medium).
2. Wet each hair with a wet sponge. Saturate them well.
3. Repeat step 1 with the wet hairs.

CONCLUSIONS:

1. Rank the hairs for length, comparing the shrinkage factor.
2. Did some hair samples hold the water longer? Did some seem to shrink more?
3. Complete the "Relative Humidity Weather Chart" for two days.

Relative Humidity Weather Chart

Date	Air Temperature	Time	% Relative Humidity	Weather Conditions

Name ______________________ Date ______________________

S6: Understanding Sedimentation

EQUIPMENT:

Clear plastic water pitcher or jar with lid
Sand (great if from the beach!) and soil
Assorted pebbles, rocks, seashells
Water
Spoon
Measuring cup

DIRECTIONS:

1. Fill the plastic pitcher with equal amounts of sand, soil, and assorted pebbles, rocks, seashells. (For example, use one cup of each.)
2. Mix them together with a spoon.
3. Add three cups of water to the pitcher. Put on the pitcher's lid and shake the pitcher well. Be sure the lid is on tight!
4. Invert the jar and allow the contents to mix well. Set aside on the counter and leave undisturbed for awhile.
5. Observe the layers of material and the settling pattern of the contents.

CONCLUSIONS:

1. Draw a picture of the sedimentation pattern. What materials are on the bottom? What comes next? Label the positions of the contents from bottom to top as they finally settled.
2. How is this similar to the natural sedimentation patterns found on the bottom of the ocean or a river?
3. Did you find that water settles in a definite pattern: rocks first, followed by pebbles, sand, and then clay?

Name ________________________________ Date ________________________

S7: Insects in the Garden

Insects can be both helpful and harmful to plant and vegetable gardens. See if you can find out which are helpful and which are harmful. Start with the following list of insects, and read and find out about each one. Then classify them by placing them in one of the categories in "My List of Insects in the Garden" below.

LIST: Ladybug, Dragonfly, Grasshopper, Snails, Aphid, Hornworm

LIST FOUR OTHER INSECTS TO INVESTIGATE:

____

____

Find out about "interplanting." This is a process by which certain plants are grown together to reduce harmful insects. An example is planting broccoli with beets, or planting marigolds and herbs (like tansy) in the garden. Do you know what planting radishes with cucumbers discourages? Beetles!

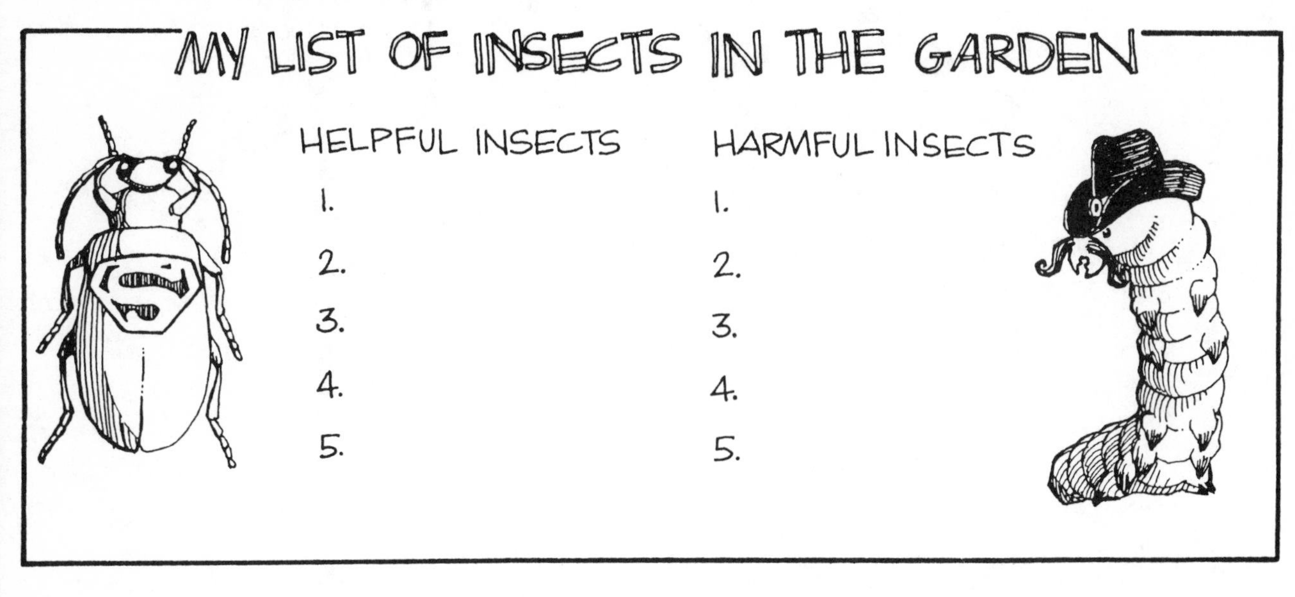

Name ______________________________ Date ______________________

S8: Examining Stems and Leaves

Look at the picture of the different patterns of leaves. Take the pictures with you outside and investigate different trees. Try to find an example of each of the six kinds:

alternate leaves	opposite
whorled	simple
needles	compound

Write the name of each kind of tree underneath the picture of each kind.

Then try to find these five parts of the stem from one sample: leaf scar, bud scar, terminal bud, lateral bud, and leaf.

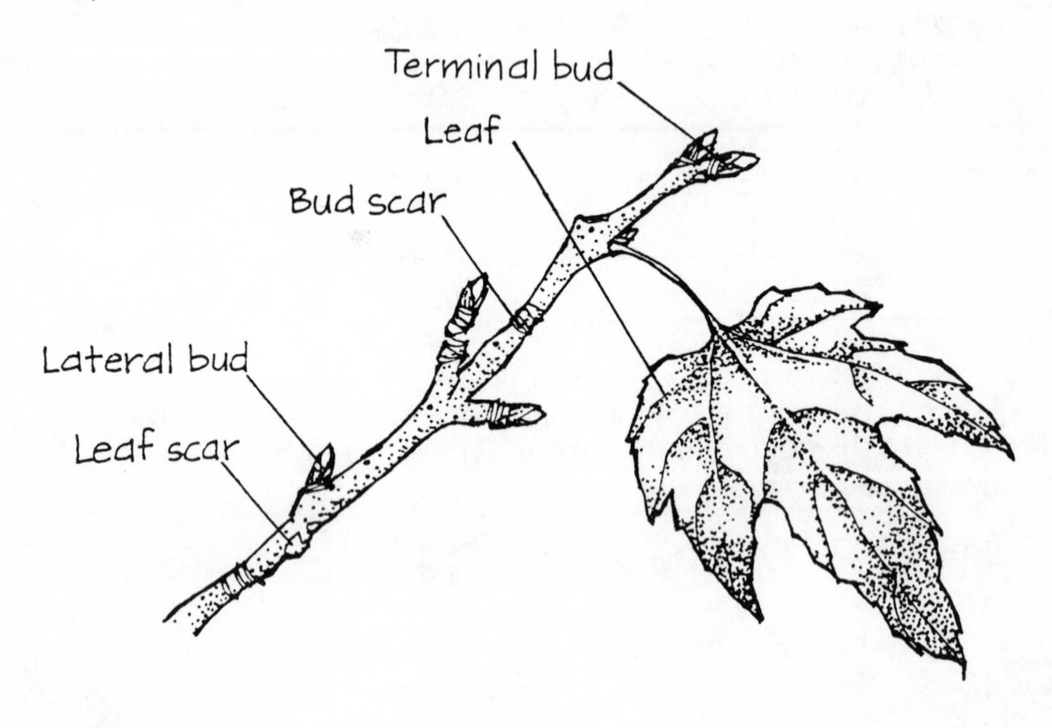

Can you also guess the age of the trees you observe?

Opposite Pattern

Alternate Pattern

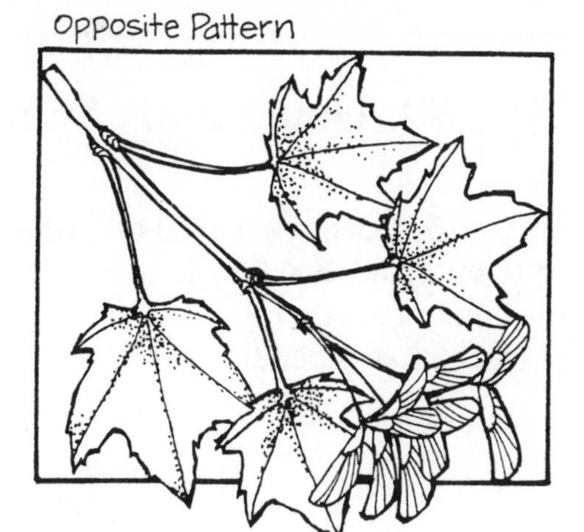

Whorled Pattern

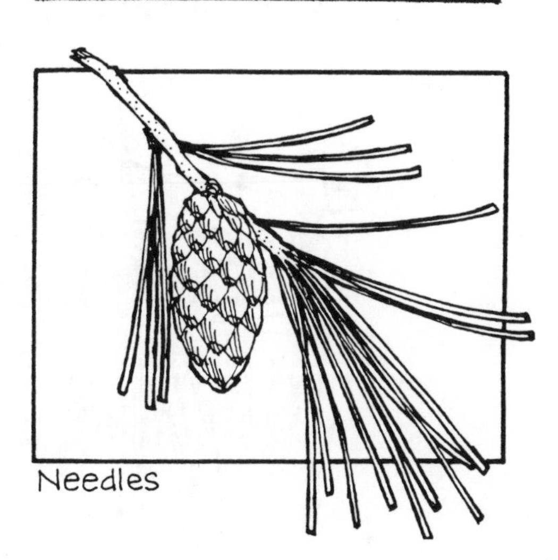

Needles

Simple Leaf

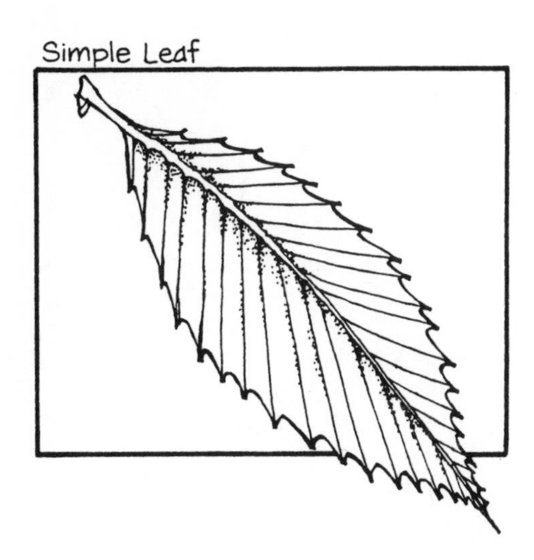

Compound Leaf

Name ______________________ Date ______________________

S9: Going on A Summer Scavenger Hunt

Using the list of 15 items, try to find or locate all of them. Use a clock to time yourself if you are locating them alone. See how long it takes you to find them.

Try this hunt on another day and see if you can beat your record. You may add or change a few items on the list depending on where you live.

SUMMER SCAVENGER HUNT LIST:

1. ROOT FROM AN ANNUAL PLANT *
2. PLANT WHORLED LEAF ARRANGEMENT *
3. EARTHWORM *
4. EARTHWORM CASTINGS
5. FERN *
6. CLOVER
7. ANNUAL PLANT *
8. PETAL OF A FLOWER *
9. TAPROOT *
10. MOTH *
11. CLOUD IN SKY (identify type) *
12. CATERPILLAR *
13. ROCK OR MINERAL (identify)
15. LADYBUG *

* LOCATE, BUT DON'T PICK OR TAKE

Name ______________________________ Date ______________________________

S10: Vegetables Grown Beneath Soil

Here is a way to grow delicious potatoes right in your own backyard! Be sure to get your parent's permission and to have a parent (or other adult) supervise the use of the paring knife.

Potatoes are tubers, a type of bulb that stores food within the round part of the plant.

EQUIPMENT:

2 certified seed potatoes (Russet, Norland)

Paring knife (CAUTION: To be used under adult supervision)

Cutting board

Spade

Area in yard or garden

DIRECTIONS:

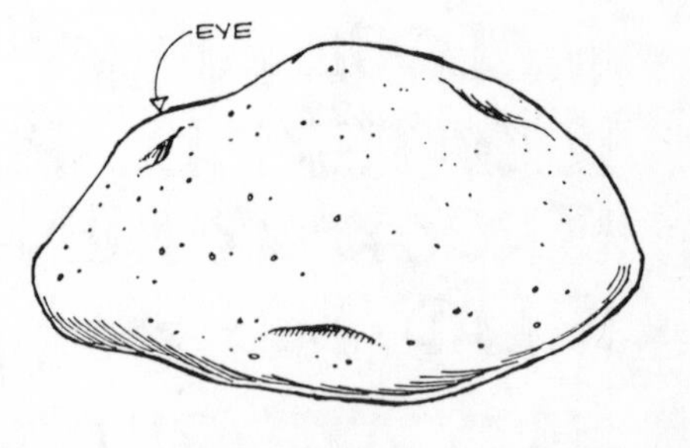

1. Find the "eyes" of each potato. The "eyes" are the little depressed parts of the potato from which the sprouts will grow. Cut each potato into four parts, so that each part has at least one eye. Be careful cutting—use the cutting board and cut slowly and carefully.
2. Plant each quarter in soil in the garden, cut-side down. Cover well with dirt. Keep it watered, but not soggy. Pull any weeds from the area, too.
3. When vines appear (in about 7 to 8 weeks after planting) and begin to turn brown, it's time to harvest your potatoes. Gently dig them out with the spade.
4. Remove your remaining potatoes from the ground prior to the first frost.

CONCLUSIONS:

What would happen if the potato was not planted beneath the soil?

Name ______________________________ **Date** ______________________

S11: The Summer Sky

Summer is a good time to view three major constellations. After viewing them, you may want to make another pattern for your constellation viewers like you did during the other seasons.

Look at the patterns of Ursa Major, Lyra, and Scorpio (also known as Scorpius). Also look at the "Star Temperature Graph." As you can see, stars are different colors, with blue-white stars being the hottest.

Lyra contains "Vega," the fourth brightest star in the sky, a white star about 20,000°F. Scorpio contains a red star, "Antares," about 5,500°F.

When you are outdoors in the summer night, look for these three beautiful constellations and think about the colors of the stars. You might want to read a "Sky Observer's Guide" and find out more about each constellation.

Then make a constellation viewer as described in activity S12.

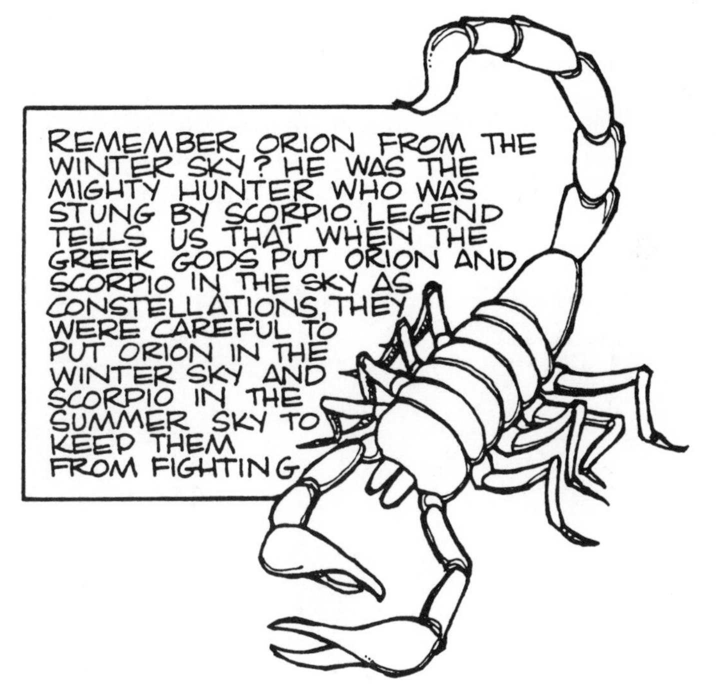

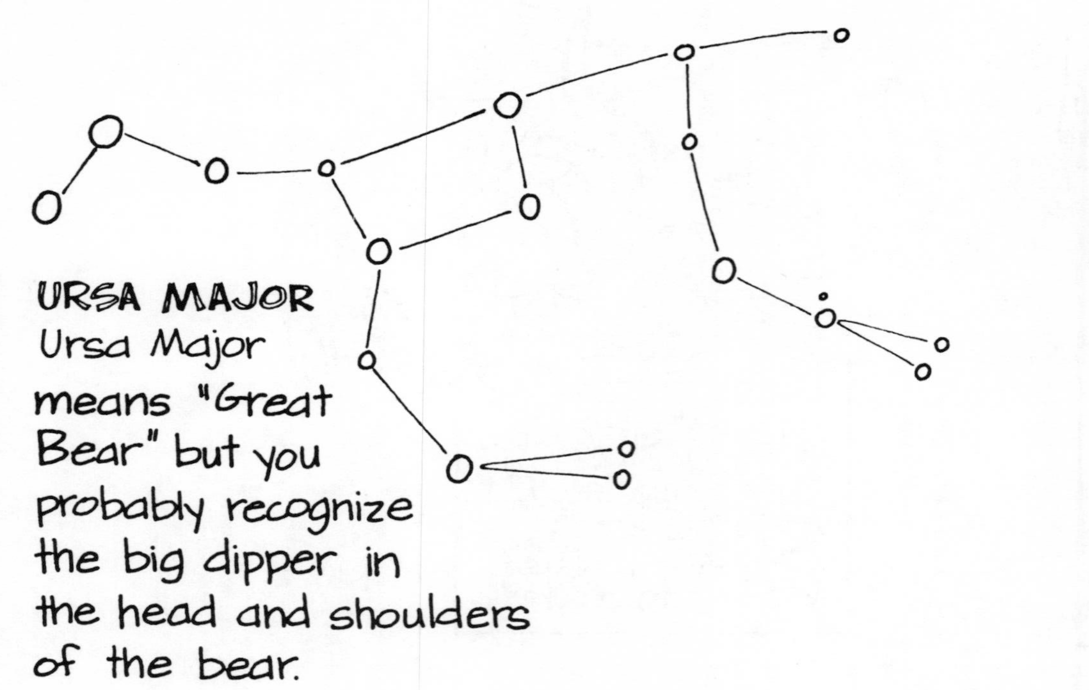

URSA MAJOR

Ursa Major means "Great Bear" but you probably recognize the big dipper in the head and shoulders of the bear.

LYRA

The Greeks saw a lyre in Lyra, but the American Indians thought it was an eagle or a vulture.

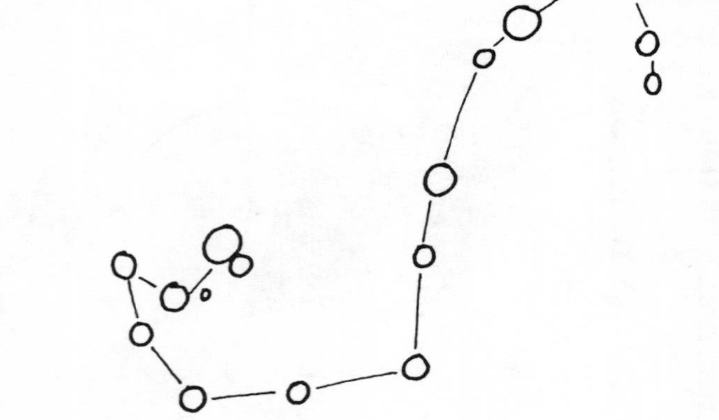

SCORPIO

Nearly every ancient culture saw the shape of a scorpion in the constellation Scorpio.

It is easy to understand how the ancient Greeks saw a bear in Ursa Major and a scorpion in Scorpio. With your pencil, make Ursa Major into a bear and Scorpio into a scorpion. Color your drawings in, if you like.

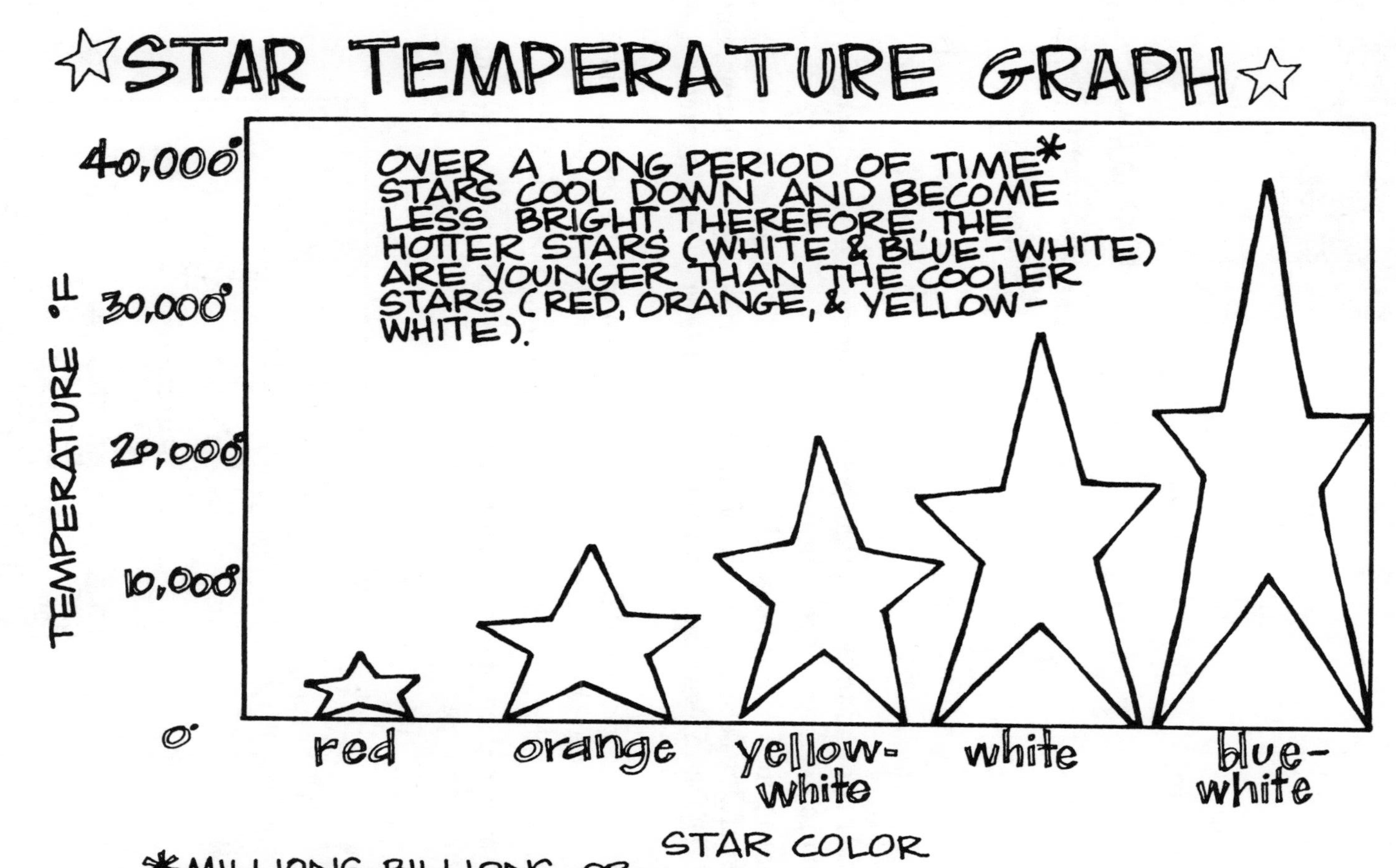

☆STAR TEMPERATURE GRAPH☆
OVER A LONG PERIOD OF TIME* STARS COOL DOWN AND BECOME LESS BRIGHT. THEREFORE, THE HOTTER STARS (WHITE & BLUE-WHITE) ARE YOUNGER THAN THE COOLER STARS (RED, ORANGE, & YELLOW-WHITE).
TEMPERATURE °F
40,000°
30,000°
20,000°
10,000°
0°
red
orange
yellow-white
white
blue-white
STAR COLOR
*MILLIONS, BILLIONS, OR EVEN TRILLIONS OF YEARS.

Name ______________________________ Date ______________________

S12: The Summer Sky—Looking for Lyra

EQUIPMENT:

Empty oatmeal box
Styrofoam cup
Patterns for Lyra
Flashlight
Markers
Sharp pencil
Scissors
Construction paper
Tape

DIRECTIONS:

1. Select the box or cup and the appropriate pattern. Cut out the paper pattern and place it on one end of your cup or box. Tape it in place.
2. Using the sharp pencil, carefully punch holes the SAME SIZE as the pattern shows. Work slowly and carefully, because the size of the holes varies; some are larger, and some are smaller. CAUTION: Always have an adult present when doing any cutting or hole punching.
3. Take the cup or box to a totally darkened room. Turn on your flashlight and shine the light inside the cup or box to shine light onto the ceiling or wall.
4. Move closer to adjust the focus of the constellation projection.
5. If you wish, you can decorate your constellation viewer on the outside with construction paper, markers, and tape.

Patterns for Lyra

Use this pattern if you are using a styrofoam cup:

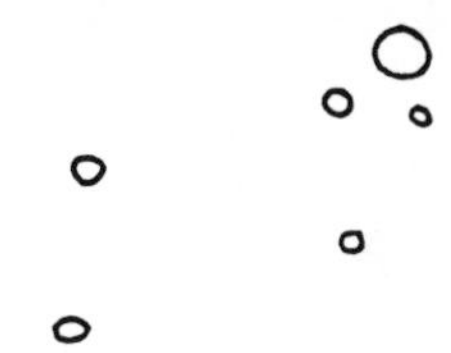

Use this pattern if you are using an oatmeal box:

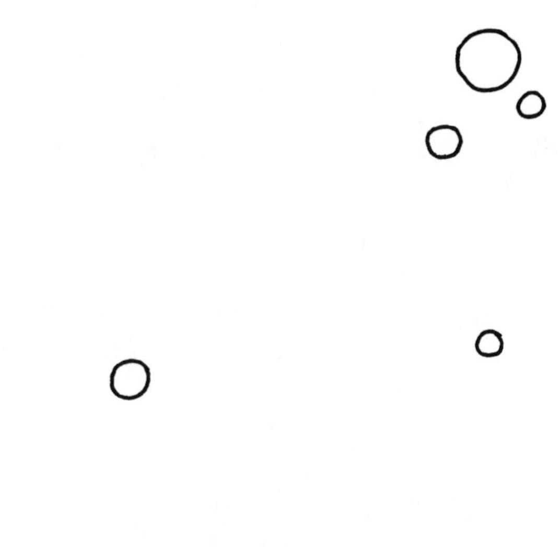

SUMMER
1. enrichment
2. activities

Name ______________________ Date ______________________

Summer Cookery—Making a Parabolic Hot Dog Cooker

Do you like to eat hot dogs? Here is a way to use the summer sun's energy to cook a hot dog!

EQUIPMENT:

- Patterns for making the cooker
- Wire coat hanger
- Wire cutter (CAUTION: To be used only by an adult)
- Plastic wrap
- Large cardboard box
- Pencil
- Aluminum foil
- Ruler
- Duct tape (found at a hardware store)
- Utility knife (CAUTION: To be used only by an adult)
- Scissors
- Plastic drinking straw
- Tape
- Hot dog (keep refrigerated until ready to cook)
- Hot dog bun, mustard, ketchup, etc.

(Special thanks to Jonathan Craig, Director of Ecology at Talcott Mountain Science Center in Avon, CT for this activity!)

DIRECTIONS:

1. Use the cardboard box for making the patterns. Be sure to make the patterns the correct sizes as shown. You will need two side panels with a focus hole on each; and one back panel.
2. Have an adult cut out the cardboard patterns with the utility knife. Bend the back panel and tabs to make a boat-shaped parabola.

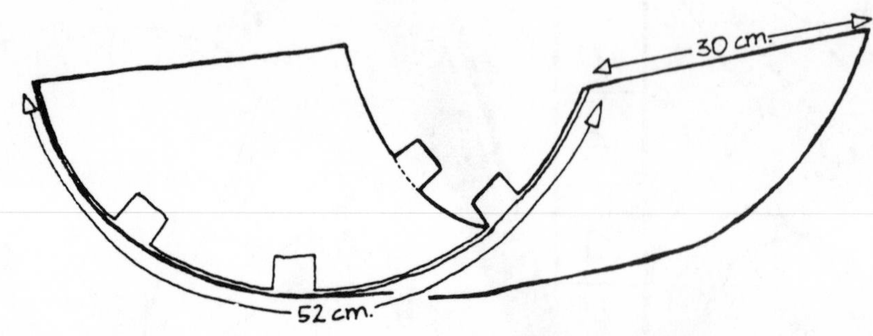

* The back panel should be bent so it matches the curve of the parabola.

3. Cover the front and back faces of the back panel and the side panels with aluminum foil. Tape, if needed.
4. Assemble the two side panels to the back panel, joining the tabs of the bottom panel to the sides of the parabola. Use duct tape to secure.

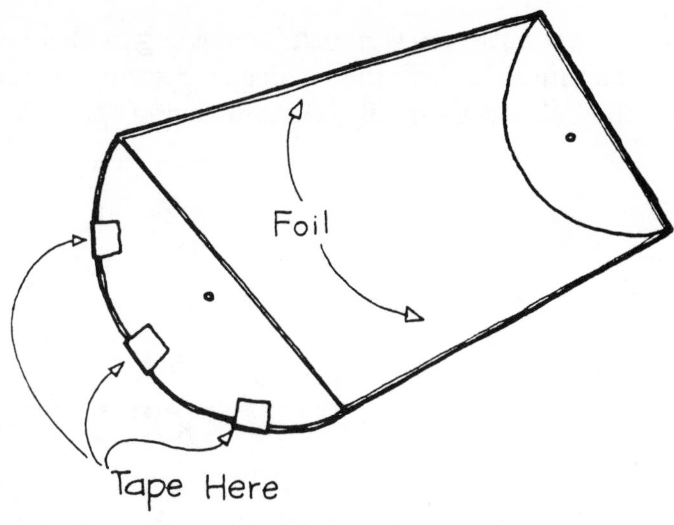

5. Have an adult cut the wire hanger to serve as a skewer/rotisserie for the hot dog. Push the hanger through the two holes of the side panels.
6. Take a straw and cut it so it extends one inch above the top of a side panel. Tape it on the exterior side of the side panel. The straw serves as an indicator for the sunlight. When cooking your hot dog, angle this straw so the sun shines through it and creates a circle of light beneath. You may want to experiment with that now.

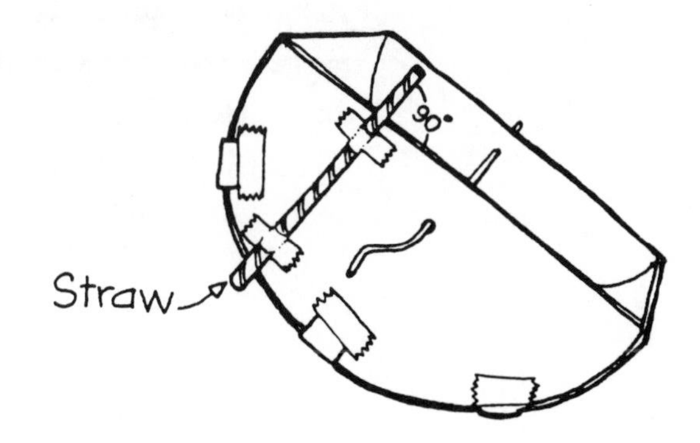

7. Remove the wire hanger and put it through the hot dog. Wrap the plastic wrap around the hot dog to cover it and seal in the juices. This also reduces the cooking time.
8. Place the wrapped hot dog on the wire hanger back into the parabolic cooker. Use the straw as an indicator and angle the straw so the circle of light shows.

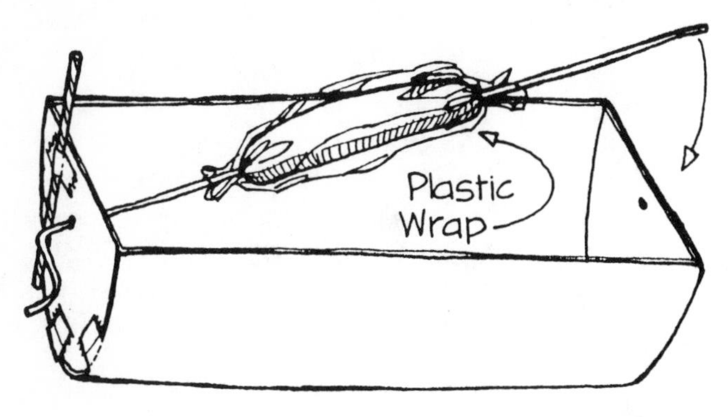

After putting the hot dog on the hanger, slip the free end back into its hole. Now you're ready to cook!

9. Although not as fast as a microwave, you can cook the hot dog in about 45 minutes or warm a pre-cooked cool one in 15 minutes. Cook the hot dog on a sunny, clear day at noon for best results. After enjoying the hot dog, find out the definition of "parabola."

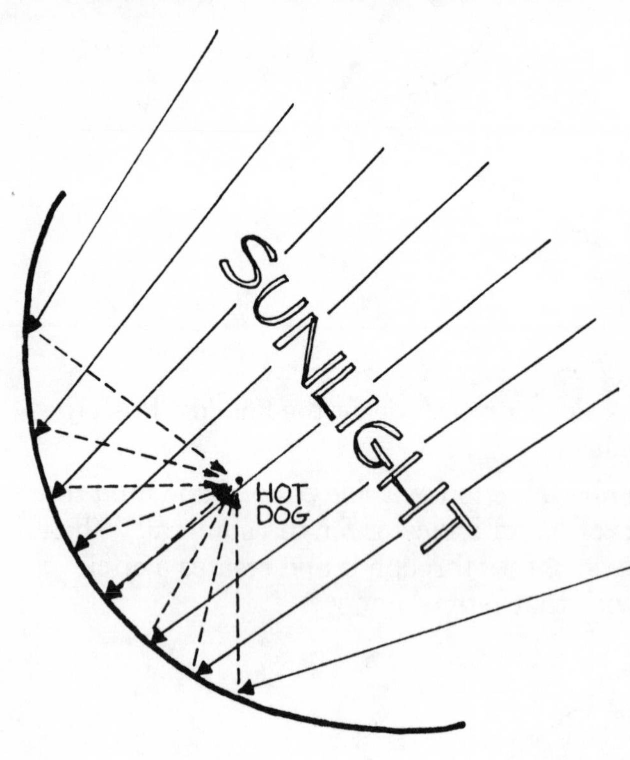

WHY DOES THIS WORK?

When sunlight hits a reflective surface (like aluminum foil) it bounces off. If that surface is shaped like a parabola (like your hot dog cooker) the reflecting rays of light are all focused on one single spot. That spot is where you have put your hot dog. So, the parabolic cooker helps the sun cook your hot dog by FOCUSING THE SUN'S RAYS ON THE HOT DOG.

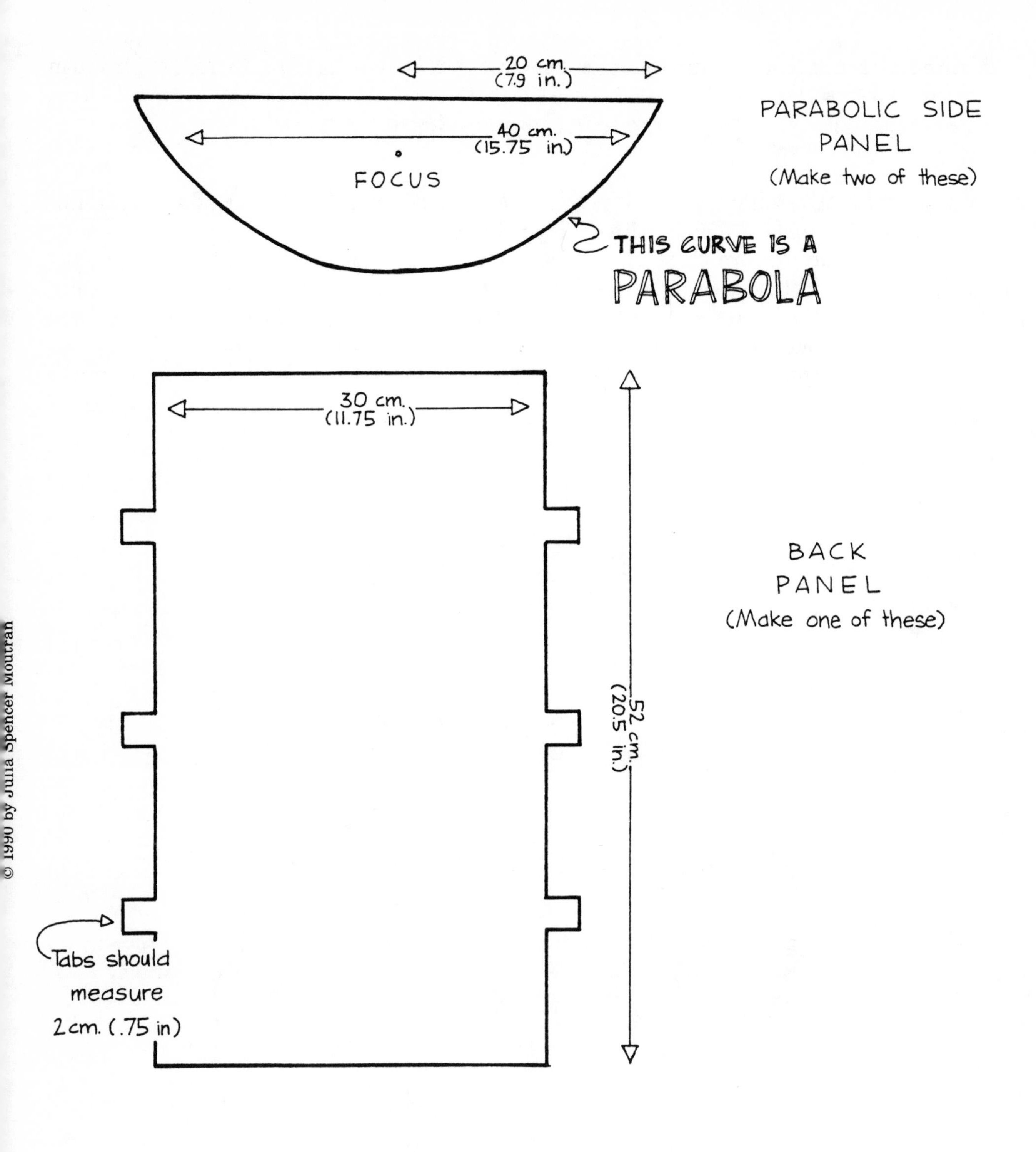
20 cm.
(7.9 in.)
40 cm.
(15.75 in.)
FOCUS
PARABOLIC SIDE PANEL
(Make two of these)
THIS CURVE IS A
PARABOLA
30 cm.
(11.75 in.)
52 cm.
(20.5 in.)
BACK PANEL
(Make one of these)
Tabs should measure 2cm. (.75 in)

Name ______________________ Date ______________________

Making Cloud Mobiles

In activity S2, "Examining Cloud Types," you found out about different cloud types. In this activity, you can make a mobile with different kinds of clouds.

Six cloud patterns are given: cirrus, altocumulus, cumulus, stratus, stratocumulus, and cumulonimbus. Write a description of each on the back of each cloud. Then use your crayons or markers to color the clouds on the other side. You might also want to use cotton that has been dipped in watercolors for more visual effects. You could add lightning, too.

Cover a hanger with yarn and suspend the six finished clouds with string from the hanger.

STRING HOLE
CUMULUS
STRING HOLE
CUMULONIMBUS
THESE CLOUDS ALMOST ALWAYS BRING THUNDER & LIGHTNING.
YOU MIGHT WANT TO MAKE SOME LIGHTNING COMING FROM THE BOTTOM OF THE CLOUD.

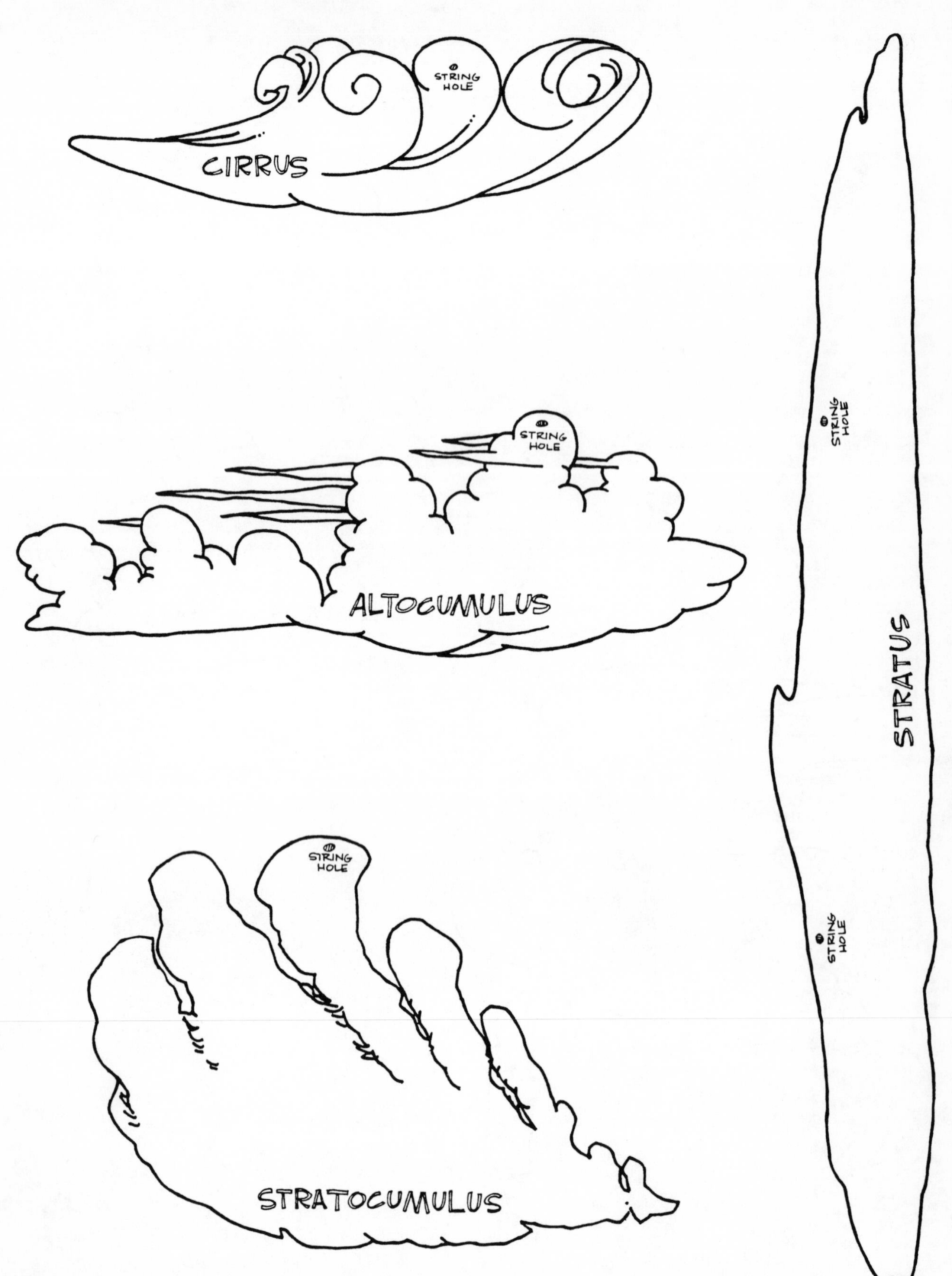
STRING HOLE
CIRRUS
STRING HOLE
ALTOCUMULUS
STRING HOLE
STRATOCUMULUS
STRING HOLE
STRATUS
STRING HOLE

Name ____________________ Date ____________________

Vegetable Printing

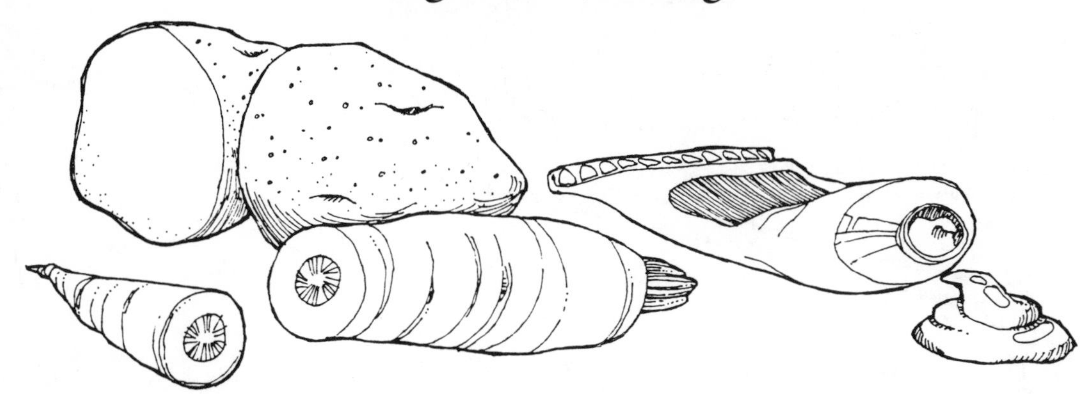

In activity S10, "Vegetables Grow Beneath Soil," you can start a garden with potatoes by planting the parts of a potato containing the "eyes" of the vegetable. In this activity, you can print the cross sections of different vegetables, such as potatoes, onions, carrots, and beans.

You may want to examine the cells of a layer of onion skin under a microscope while you have the onion available. This will help you understand the cellular structure of vegetables and plants.

EQUIPMENT:

Assorted fresh vegetables (potatoes, carrots, onions, radishes, beans, etc.)

Ink stamp pad

Tempera paint and brushes

Water

Scissors

Paring knife (CAUTION: To be used only with adult supervision)

Paper

Paper towels

DIRECTIONS:

1. Use scissors to cut paper to size for the vegetable print(s).
2. Rinse and dry the vegetables to clean them as much as possible.
3. Carefully cut each vegetable to create patterns. You might slice it crosswise, down the middle, or in different ways.
4. Apply color by using an ink pad or apply tempera paint with a brush.
5. Turn the painted side of the vegetable down onto the paper. Press lightly and remove the vegetable.
6. Continue step 5 with different colors and vegetables.
7. Allow the prints to dry before displaying. Clean up the supplies and work area.

Name ____________________

Date ________________

SUMMER FUN: CONTRASTS & COMPARISONS

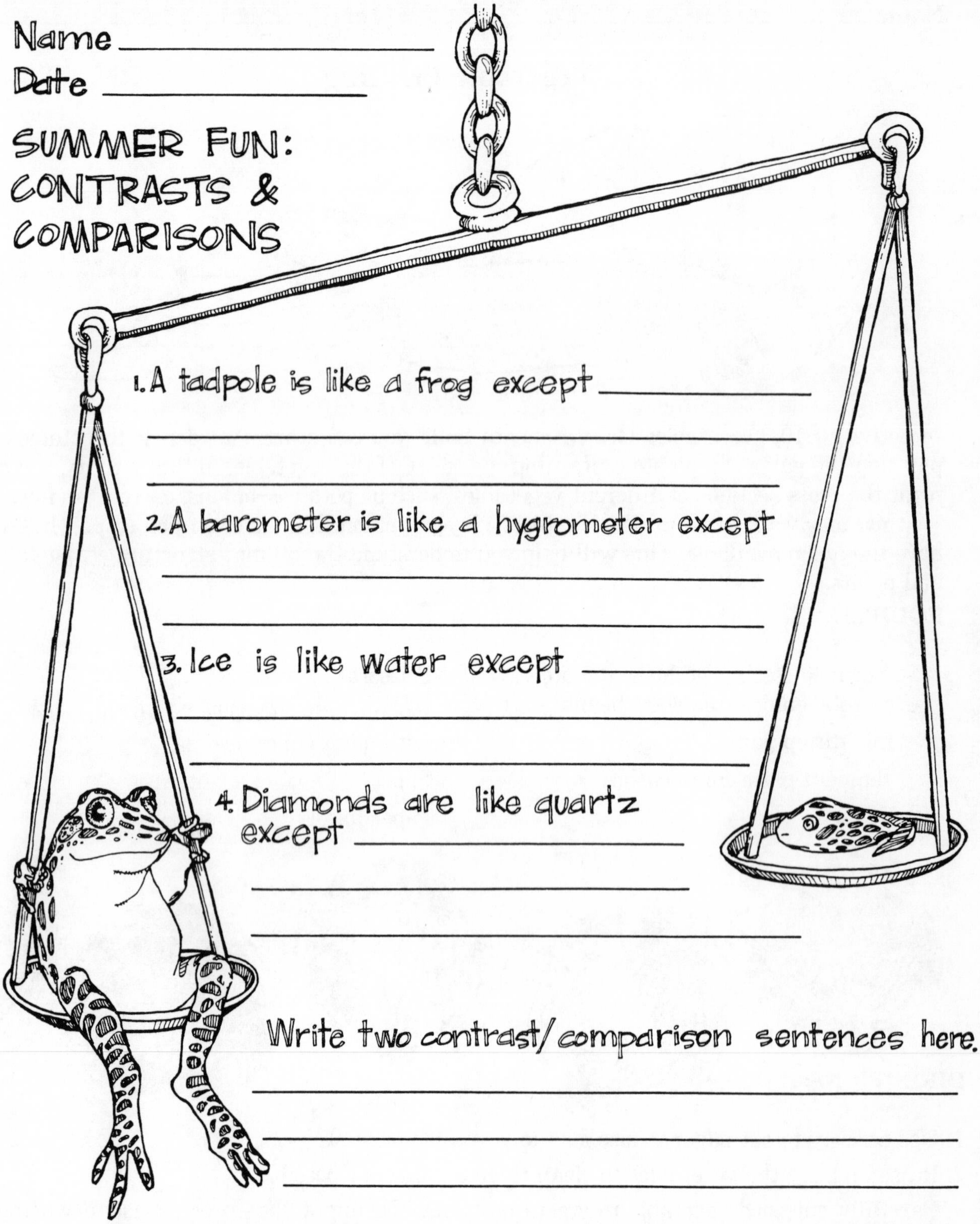

1. A tadpole is like a frog except ____________
____
____

2. A barometer is like a hygrometer except
____
____

3. Ice is like water except ____________
____
____

4. Diamonds are like quartz except ____________
____
____

Write two contrast/comparison sentences here.
____
____
____
____
____
____

SUMMER SCIENCE SONG

Music by: Patricia McCamish Donohoe Lyrics by: Julia Moutran

"MY FAVORITE SEASON"

Name ______________________ Date ______________________

ACROSS:

2. The fourth brightest star in the sky, with a temperature of 20,000°F
3. Small, elongated, soft bodied animal that moves through the soil
4. The action or process of settling or sediment depositing
6. A plane curve having an equal distance from a fixed point and a fixed line

DOWN

1. A weather instrument designed to measure relative humidity.
3. A colorless liquid consisting of two parts hydrogen and one part oxygen
5. A word that describes an environment deficient or low in moisture
7. The wormlike larva of a butterfly or moth

(NOTE: A new word you will need is "xeric," pronounced zir-ik.)

Name ______________________ Date ______________________

Summer Herbs to Identify

Can you guess these popular summer herbs? They are often found in the garden and used in cooking to enhance the flavor of foods.

I'm most famous for the flavor I add to pickles.

I AM ______________________

My leaves are dried and chopped to be used in tomato sauce and many Italian dishes.

I AM ______________________

Although it is spelled differently, my name sounds like what you would find if you looked at your watch.

I AM ______________________

I have a girl's name. I am delicious in soups and stews.

I AM ______________________

You might have bread flavored with me in an Italian restaurant.

I AM ______________________

My curly green leaves are chopped and sprinkled over many foods. Or you might see just a sprig of me on the side of your plate.

I AM ______________________

Name ________________________ Date ________________________

Creating a Shell Cast or Mold

Perhaps you have seashells at home or at the beach you can use for this. If not, you can create the molds using small pebbles and rocks.

EQUIPMENT:

- Bucket or pan
- Shells or rocks
- Modeling clay
- Plaster of paris
- Petroleum jelly
- Spoon
- Toothpick
- Acrylic or tempera paint

DIRECTIONS:

1. Line the bucket or pan with the modeling clay.
2. Spread petroleum jelly on top of the clay and on the sides of the container.
3. Press shells or rocks into the clay. Then lift them out with a toothpick and spoon so that their impressions are left in the clay.
4. Mix the plaster of paris with water according to the directions on the package. Pour it into the container. It will get warm as it sets.
5. Let the plaster cool and then remove it from the container.
6. Paint your plaster mold with tempera or acrylic paint when it is thoroughly dry.

THINK:

What kind of animals once lived in your seashells? Read and find out about them. If you used rocks, can you identify them?

Name ______________________________ Date ______________________

Summer Photography

Here is a good project for you during the summer. When you return to school in the fall, you can bring the finished project with you.

All you need is film, a camera, posterboard, glue or tape, and markers.

Photograph many of the wonders of nature—scenes from your observations of trees, stems, and leaves…the Scavenger Hunt…birds eating, swimming, and nesting…flowering plants and gardens…insects. The list is really endless. You might also think about your summer science activities and photograph a few of the finished products.

Display all of these photos. If you learned something special about certain activities, write a few sentences about them and make a label or photograph caption to go beneath or on the back of the display poster. You might also write your facts on index cards and keep them to share with others who will be interested in your display. In this way, you'll be teaching others about the amazing world of science and nature!

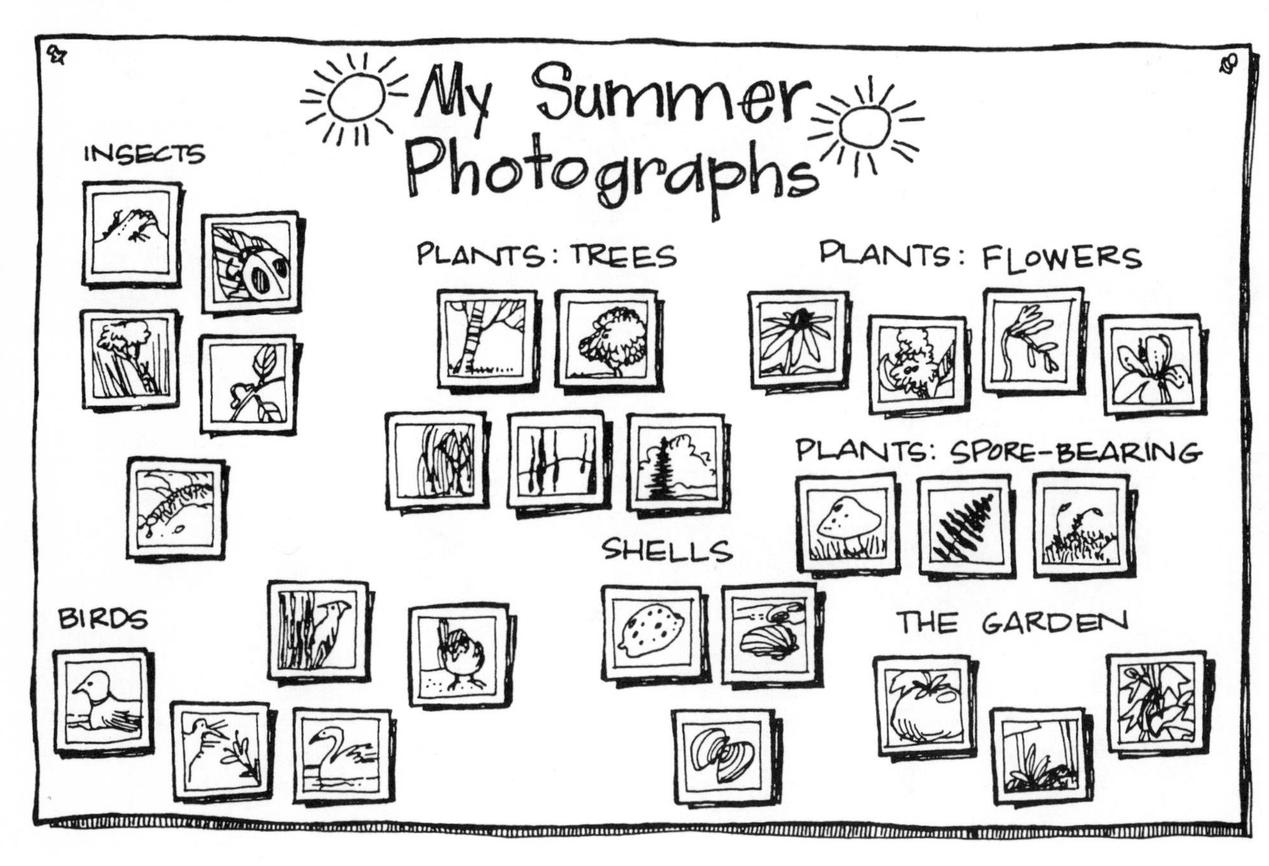

APPENDICES

1: ANSWER KEY
2: ORDERING INFORMATION FOR MATERIALS
3: CORRELATED SCIENCE VIDEOS AND 16mm SCIENCE FILMS

Appendix 1: ANSWER KEY

F4: LOOKING FOR SIGNS OF PHOTOSYNTHESIS

ANSWERS:

SAMPLE	PRESENCE OF CHLOROPHYLL	PRESENCE OF SPORES	MANUFACTURE OWN FOOD
MUSHROOM	NO	YES	NO
FERN	YES	YES	YES
LEAF	YES	NO	YES
ELODEA	YES	NO	YES

FALL CROSSWORD PUZZLE I

Across
2. POLE
4. DECIDUOUS
6. BULB
8. MAGNET
9. NOUN
11. FALL

Down
1. COMPLEX
3. SIMPLE
5. EQUINOX
7. HAIKU
10. PUMPKIN
12. HARVEST

DIFFERENT LEAVES: UNSCRAMBLE ME!

1. RED MAPLE
2. ASPEN
3. BIRCH
4. OAK

FALL CROSSWORD PUZZLE II

Across
1. EQUINOX
3. HEMISPHERE
4. AUTUMN
5. SUN
8. DAY

Down
2. AXIS
3. HOURS
6. NIGHT
7. EQUATOR

WII: EXAMINING WINTER WEATHER

1. St. Pete/Tampa
2. Milwaukee
3. Austin, Oklahoma City
4. Atlanta, Buffalo, Los Angeles
5. February

WINTER CROSSWORD PUZZLE

Across
2. AURORA
5. GEOTROPISM
7. DIFFUSE

Down
1. WINTER
3. RIGEL
4. SOLSTICE
6. TAPROOT
8. CETUS

COOL- AND WARM-WEATHER CROPS

Cool
1. RADISH
2. CARROT
3. POTATO
4. PEAS
5. PARSLEY
6. OKRA
7. ONION
8. BRUSSELS SPROUTS
9. CAULIFLOWER
10. BROCCOLI
11. SPINACH
12. BEET
13. CELERY
14. LETTUCE

Warm
1. TOMATO
2. PEPPER
3. BEANS.
4. CUCUMBER
5. STRING BEANS
6. EGGPLANT
7. CORN
8. AVOCADO
9. ZUCCHINI
10. WATERMELON

SPRING SCIENCE STORIES, SONGS, & POETRY: DECODE ME!

LYRA
ORION
PEGASUS
CANCER
TAURUS
LEO
LIBRA
CYGNUS
ARIES
CETUS

SPRING CROSSWORD PUZZLE I

Across

2. MOLT
3. MONARCH
5. NYMPH
8. LARVA
9. THREE
10. FALL
11. WINTER

Down

1. COCCOON
4. CHRYSALIS
6. SPRING
7. CATERPILLAR

SPRING CROSSWORD PUZZLE II

Across

1. EQUINOX
3. FRESNEL
5. GRAVITY
8. COMPOST

Down

1. BROCCOLI
4. SEPAL
6. NYMPH
7. ORTHOPTERA

SUMMER CROSSWORD PUZZLE

Across

2. VEGA
3. WORM
4. SEDIMENTATION
6. PARABOLA

Down

1. HYGROMETER
3. WATER
5. XERIC
7. CATERPILLAR

SUMMER HERBS TO IDENTIFY

dill
thyme
garlic
basil
rosemary
parsley

Appendix 2: ORDERING INFORMATION FOR MATERIALS

GLASS PEG MOUNTS

These can be ordered from:

Wess Plastic
50 Schmitt Boulevard
Farmingdale, NY 11735-1484
(516) 293-8944

The Glass Peg Registered Slide Mounts used in the four constellation lessons can be ordered inexpensively from Wess Plastic. Ask for #2 Glass Peg Registered Slide Mounts. (Catalog #849; inserts and price list #849-P3)

CORRELATED VIDEOS AND FILMS

All correlated videos and films (see Appendix 3) may be ordered directly from:

CORONET/MTI Film & Video
108 Wilmot Road
Deerfield, IL 60015
(312) 940-1260
1-800-621-2131

The videos and films listed in Appendix 3 are optional. Each has been reviewed and is appropriate for the intermediate grade levels. (Consult the latest catalog for price and rental information.)

SCIENCE EQUIPMENT AND MATERIALS

Most of the equipment, live specimens, prepared microscopic slides, aquarium or plant cultures, and seeds and bulbs can be ordered from:

Connecticut Valley Biological Supply Co.
P.O. Box 326
82 Valley Road
Southampton, MA 01073
1-800-628-7748
(413) 527-4030

Call or write for a complete catalog. All chemicals needed for the experiments in this book can also be ordered from this company.

BOOKS

These books by Julia Spencer Moutran are recommended to accompany several activities. They may be ordered from your local bookstore, through Baker and Taylor Distributors, or through the publishers:

Collecting Bugs and Things
ISBN: 0-8431-2226-9
Published by: Price/Stern/Sloan
360 North LaCienega Boulevard
Los Angeles, CA 90048
1-800-227-8801

The Story of Punxsutawney Phil, the Fearless Forecaster
ISBN: 0-9617819-2-0 (hardcover)
ISBN: 0-9617819-0-4 (paper)
ISBN: 0-9617819-3-9 (audiocassette)
Published by: Literary Publications
34 Oak Bluff
Avon, CT 06001

Appendix 3: CORRELATED SCIENCE VIDEOS AND 16mm SCIENCE FILMS

The following videos and 16mm films correlate with the activities in *Science Activities for All Seasons* and may be ordered for rental or purchase through Coronet/MTI.

Some of the videos and films relate to more than one activity or season, so consult the list for appropriate titles and read the descriptions. You may want to share the videos/films with other teachers on your grade level and correlate the activity sheets with the films/videos.

All of the videos and films have been previewed for this book. Each one will help enrich and extend the science concepts presented.

NOTE: Grade levels' key is P = Primary (K-3), I = Intermediate (4-6), J = Junior High, H = High School, A = Adult, and C = College.

TITLE & DESCRIPTION	FALL	WINTER	SPRING	SUMMER
Autumn Comes to the Forest As the days grow cooler, the animal activities of summer slowly change. Birds begin to migrate, and other animals gather and store food. A shimmering cascade of colored leaves signals the essence of the autumn change. Fattening mammals prepare for hibernation as cold winds are followed by the first snow. *Grade level:* P, I *Time:* 11 minutes *Catalog #:* 1964 (16mm) 1964V (video)	F1 F2 F3 F4 F7			
The Bird's Year A fascinating look at the noticeable differences among even the most common species of birds in the way they conduct their lives through the course of a year. Differences in food requirements, migration, nest building and the feeding of their young are explored. Are all birds alike? This film provides surprising answers. (International Wildlife Film Festival award winner) *Grade level:* I, J *Time:* 12 minutes *Catalog #:* SS242 (16mm/video)	Any time during Fall	Any time during Winter	Any time during Spring	Any time during Summer
Butterflies An amazing look at the beauty and variety of butterflies and their unique color patterns, temperature regulation, and wing forms. *Grade level:* I, J, A *Time:* 13 minutes *Catalog #:* SS229 (16mm/video)			Any time during Spring	Any time during Summer

TITLE & DESCRIPTION	FALL	WINTER	SPRING	SUMMER
Caterpillars The immature forms of moths and butterflies have two jobs: eat and grow. How does the caterpillar protect itself against predators during the winter? This film examines some surprising methods including camouflage, mimicry, and protective coloring. *Grade level:* I, J, A *Time:* 13 minutes *Catalog #:* SS224 (16mm video)		Any time during Winter		
The Cell—Structural Unit of Life (Second Edition) Exploring the wonders of life through a microscope, young biologists discover that plants and animals are all composed of living cells. Through photomicrography, they explore the nucleus, cytoplasm, and cell wall that all together are called "protoplasm." They witness growth and reproduction in single-celled and multi-celled organisms. *Grade level:* I, J *Time:* 11 minutes *Catalog #:* 3806 (16mm) 3806V (video)	F5 F6			
Clouds and Precipitation From dazzling white to ominous black, from ground-hugging mists to towering thunderheads, clouds bring beauty to the skies and water to the land. Time lapse studies show how clouds form and vanish into the air, while satellite, aerial, and ground-based photography conveys basic information about cumulus, stratus, and cirrus clouds. *Grade level:* I, J, H, C *Time:* 15 minutes *Catalog #:* 4623C (16mm/video)	F8 F11	W2 W5 W6		S2
Crystals and Their Growth The transparent brilliance of a quartz crystal, the jewel-like patterns of snowflakes and salt crystals growing before students' eyes—these help illustrate the nature of crystals and how they grow. Methods are shown for growing several types of crystals. Scenes of industrial uses of crystals indicate their		W6		

TITLE & DESCRIPTION	FALL	WINTER	SPRING	SUMMER
importance in present-day science and technology. *Grade level:* J, H *Time:* 12 minutes *Catalog #:* 1717 (16mm) 1717V (video)				
Development (Discovering Insects) Witness the dramatic process of insect metamorphosis. *Grade level:* I, J, A *Time:* 13 minutes *Catalog #:* SS227 (16mm/video)				S7 S9
Discovering Insects Series Part of the award-winning "Many Worlds of Nature," this series uses close-up photography to capture the world of these most abundant and least understood creatures on Earth. Produced by Morse-Allen, Inc. *Grade level:* I, J, A *Time:* 13 minutes *Catalog #:* SS227 (16mm/video)			Any time during Spring	Any time during Summer
The Earth: Its Water Cycle Rolling cumulus clouds rapidly mass in time-lapse speed and then quickly dissolve to invisible water vapor; ice crystals grow under photomicrography; and in a plastic box, air and water under varying pressures and temperatures help clarify the evaporation/condensation/precipitation science concepts that underlie the water cycle of the earth. *Grade level:* J, H *Time:* 11 minutes *Catalog #:* 3519 (16mm) 3519V (video)	F5 F8	W5 W6		S2 S4 S5
The Earth's Atmosphere A breathtaking journey upward through the atmosphere with a mountain climber, research balloon, and passenger jet reveals our remarkable planet and the atmosphere's layered structure produced by gravity and interactions with radiation from the sun. *Grade level:* I, J, H, C *Time:* 14 minutes *Catalog #:* 4620C (16mm video)		W4 W9		

TITLE & DESCRIPTION	FALL	WINTER	SPRING	SUMMER
Electricity and Magnetism Many devices that power technological civilization depend on a relationship between motion and two basic properties of matter—electricity and magnetism. Animated demonstrations show how forces exerted by electric charges and magnets are related and how their interactions cause motion Applications are shown in motors, generators, and transformers. Their relationship to Earth's magnetic field, the Borealis, and nuclear fusion is explored. *Grade level:* I, J, H *Time:* 17 minutes *Catalog #:* 4602 (16mm) 4602V (video)		W9		
Evergreens An in-depth look at evergreens and the reason why these hearty trees do not lose their leaves like most other trees. *Grade level:* I, J *Time:* 12 minutes *Catalog #:* SS255 (16mm/video)		W3		
Flowers The role of flowers in nature, and their relationship to insects. *Grade level:* I, J *Time:* 12 minutes *Catalog #:* SS256 (16mm/video)			SP4 SP4 SP5	
Food: Energy for Life Time-lapse photography, photomicrography, and extreme close-up shots form a beautiful photographic essay on the eternal cycle of life. The important concepts of photosynthesis, food chains, and recycling of nutrients through death and decay are shown to be vital to the balanced, unending cycle of nature. *Grade level:* J, H *Time:* 10 minutes *Catalog #:* 3852 (16mm) 3852V (video)	F2 F3 F4			
Garden Plants and How They Grow (Revised) This adventure into the microcosm of the garden explores various ways in which a plant			SP2	S7 S10

TITLE & DESCRIPTION	FALL	WINTER	SPRING	SUMMER
starts to grow and what happens inside the plant as it grows. Youngsters see where the plant's food comes from, how it is made, and who the plant's friends and enemies are. This is an excellent guide for young gardeners (and garden watchers). *Grade level:* P, I *Time:* 11 minutes *Catalog #:* 4399 (16mm) 4399V (video)				
Global Forecasting How are weather forecasts made? Where does the information come from? This program examines the global network of radar stations—hundreds of remote spots as well as weather satellites in orbit—from which meteorologists form the basis for their increasingly accurate weather reports. *Grade level:* I, J, H, C *Time:* 14 minutes *Catalog #:* 4626C (16mm video)	F11		SP7	S4 S5
Growth of Flowers (Second Edition, Revised) Through the magic of time-lapse photography, the growth of flowers becomes a dance of beauty choreographed by the shifting sun, changing temperatures, and the flower's own patterns of development. Viewers see how flowering plants live through the day, a season, and a year; how insects help them reproduce; how each plant responds in its own unique way to the changing environment. A kaleidoscope of colors and forms becomes the "miracle of the flowers." *Grade level:* P, I, J *Time:* 11 minutes *Catalog #:* 4400 (16mm) 4400V (video)			SP2 SP4 SP5	
How Green Plants Make and Use Food (Revised) All living things on this planet depend, directly or indirectly, on green plants for food and oxygen. Using clear animation and time-lapse micrography and microfootage, this film shows how materials reach the leaves and how light supplies energy for the manufacture of sugar. This film clarifies the difference between		W8	SP2	

TITLE & DESCRIPTION	FALL	WINTER	SPRING	SUMMER
food and raw materials; between photosynthesis and cellular respiration. *Grade level:* I, J *Time:* 12 minutes *Catalog #:* 4379 (16mm) 4379V (video)				
Insect Life Cycles Close-up photography provides stunning examples of both simple and complete insect metamorphosis manifested in the praying mantis, cecropia moth, polyphemus moth, various butterflies, mud dauber wasp, dragonfly, and others. An unusual segment of the film depicts the life cycle of the periodic cicada. A Centron film. (Golden Eagle, CINE award winner) *Grade level:* I, J, H *Time:* 15 minutes *Catalog #:* 80524 (16mm) 80524V (video)			Any time during Spring	Any time during Summer
The Monarch and the Milkweed Explore the close and very special relationship between the milkweed plant and the monarch butterfly. *Grade level:* I, J *Time:* 12 minutes *Catalog #:* SS247 (16mm/video)			SP4	
The Oak No other wild plant in North America provides food and shelter for more wild animals than the oak. *Grade level:* I, J *Time:* 12 minutes *Catalog #:* SS251 (16mm/video)	F2		SP5 SP6	
Of Birds, Beaks and Behavior The stucture of a bird's beak affects its habitat, diet, behavior and other aspects of its appearance. *Grade level:* I, J *Time:* 12 minutes *Catalog #:* SS248 (16mm/video)		Any time during Winter	Any time during Spring	
Pollination Mechanisms An in-depth look at cross-pollination and self-pollination. Picks up where the film *Flowers* left off.			SP2 SP4 SP5 SP6	

TITLE & DESCRIPTION	FALL	WINTER	SPRING	SUMMER
Grade level: I, J *Time:* 12 minutes *Catalog #:* SS257 (16mm/video)				
Seed Dispersal How plants disperse their seeds in order to reproduce. *Grade level:* I, J *Time:* 12 minutes *Catalog #:* SS253 (16mm/video)	F2 F9			
Seeds Grow Into Plants (Revised) Seeds grow on plants and plants grow from seeds. It's a process that can be seen about us in trees, flowers, grass, and even weeds. This film follows the complete cycle of nature—showing types of seeds, where they travel, and, through time-lapse photography, how they sprout and grow into plants like those from which they came. *Grade level:* P, I *Time:* 7 minutes *Catalog #:* 3941 (16mm) 3941V (video)	Any time during Fall		Any time during Spring	
Seeds: How They Germinate Learning about seeds and how they germinate tells a lot about the success of seed plants. Students see the system of seed dispersal, the structure of the embryo, the difference between monocotyledon and dicotyledon and the stages in the growth of the embryo, seedling, and plant. *Grade level:* I, J *Time:* 11 minutes *Catalog #:* 3111 (16mm) 3111V (video)	F4			
Spring Comes to the Forest The white stillness of winter gives way to melting brooks, uncurling new leaves, and the sound of returning birds. As the sun warms the earth, the forest is vivid with spring flowers. Animals find mates and raise their young as the warm spring days lengthen toward summer. *Grade level:* P, I *Time:* 10 minutes *Catalog #:* 1963 (16mm) 1963V (video)			SP1 SP4 SP5 SP6	

TITLE & DESCRIPTION	FALL	WINTER	SPRING	SUMMER
Surviving the Cold Some ways that animals, birds, and insects deal with the problems of winter. *Grade level:* I, J *Time:* 12 minutes *Catalog #:* SS254 (16mm/video)		W4		
Tree Blossoms Best known for their leaves, trees are flowering plants of great beauty. *Grade level:* I, J *Time:* 12 minutes *Catalog #:* SS258 (16mm/video)			SP4 SP5 SP6	S8
Trees: Their Flowers and Seeds Shows a great variety of tree flowers and differences in complete and incomplete flowers. Extreme close-ups and animation illustrate petals, sepals, stamens, and pistils, and show how flowers are pollinated to produce fruits containing the seeds for another generation. *Grade level:* P, I, J *Time:* 10 minutes *Catalog #:* 1907 (16mm) 1907V (video)			SP4 SP5	S8
Understanding Our Earth: Soil (Revised) A soil scientist presents a lucid explanation of the soil profile (topsoil, subsoil, loose rock, bedrock) plus the elements of soil. The process of soil formation is shown—rocks breaking down by erosion and decaying matter changing into humus. Viewers see factors that make soils differ and the importance of conservation. (Part of "Understanding Our Earth" series. *Grade level:* I, J *Time:* 12 minutes *Catalog #:* 3845 (16mm) 3845V (video)	F5			S6 S10
Violent Storms Dramatic cinematography and time-lapse photography convey the drama of nature's most spectacular display—the thunderstorm. This program follows atmospheric scientists as they study the mysteries of thunderheads. Technological advances such as satellites and Doppler radar offer the promise of spotting developing tornadoes before they can be seen from land.				S2

TITLE & DESCRIPTION	FALL	WINTER	SPRING	SUMMER
Grade level: I, J, H, C *Time:* 14 minutes *Catalog #:* 4625C (16mm video)				
Weather Systems in Motion How do the great mid-latitude weather systems get started and how does their movement account for alternating stormy and fair weather? This film clarifies the movement of air masses, compares cyclones and anticyclones, and shows how waves in the upper atmosphere affect the development of warm and cold fronts. *Grade level:* I, J, H, C *Time:* 14 minutes *Catalog #:* 4624C (16mm video)	F11			
What Do Seeds Need to Sprout? Simple experiments are performed to find out what things help a seed to germinate. Comparisons of visual evidence in time-lapse photography are made and commented on by a boy and girl. They reach three conclusions: the right temperature, water, and air are needed for seeds to sprout. *Grade level:* P *Time:* 11 minutes *Catalog #:* 3432 (16mm) 3432V (video)	F4		SP2	
Winter Comes to the Forest When winter comes, the forest looks cold, bare, and dead. But, if anyone looks more closely, there are buds containing tiny leaves, the coccoon of a moth, a chickadee looking for insect eggs, rabbit tracks in the snow, or seeds that will grow next spring. These reveal the many ways that living things are adapted for living through the winter. *Grade level:* P, I *Time:* 11 minutes *Catalog #:* 1395 (16mm) 1395V (video)		W2 W10		
Winter Signs Learn to identify various structures, marks, and other signs made by insects during winter. *Grade level:* I, J, A *Time:* 13 minutes *Catalog #:* SS231 (16mm/video)		W3 W10		

Notes